Termites and heritage buildings

Termites and heritage buildings

A study in integrated pest management

Brian Ridout

Historic England

Published by Liverpool University Press on behalf of Historic England, The Engine House, Fire Fly Avenue, Swindon SN2 2EH
www.HistoricEngland.org.uk

Historic England is a Government service championing England's heritage and giving expert, constructive advice.

The views contained in this book are those of the author alone and not Historic England or Liverpool University Press.

First published 2023

ISBN: 978-1-80207-839-8 hardback

British Library Cataloguing in Publication data
A CIP catalogue record for this book is available from the British Library.

Brian Ridout has asserted the right to be identified as the author of this book in accordance with the Copyright, Designs and Patents Act 1988.

Typeset in Charter 9/11

Page layout by Carnegie Book Production

Printed in the UK by Gomer

Front cover: Worker subterranean termites (*Reticulitermes balkanensis*) in Skopelos, Greece. Reproduced with permission from the photographer, Elizabeth Ridout.

Contents

Preface

It seems inevitable, as the climate becomes warmer, that termites will become an increasing threat to UK heritage buildings. However, for the foreseeable future, the land is unlikely to be overrun with them as found in many tropical countries, and the development or importation of a dedicated eradication industry will not be necessary. Those concerned with heritage buildings should be able to contain the problem if they have a basic understanding of it.

This book is intended to provide background information on pest insects that have previously been unfamiliar in the UK. It inevitably contains scientific names and unfamiliar terminology, but every effort has been made to explain all this and make the text accessible to a wide audience. It also introduces the remedial methods that have been devised worldwide so that the least destructive and most appropriate can be selected within the context of minimum intervention and the conservation of heritage structures.

It is becoming increasingly apparent that our dependence on pesticide formulations is misplaced. The concomitant concept of integrated pest management – the combination of pesticides with an understanding and modification of pest behaviour and environmental requirements – has been applied to termites. However, it has been seen as an alternative for the pesticide industry. Liquid termiticides and baits are both commercially available and offered with the basic statement that the building should be kept dry and free from cellulose-based debris. This book, while discussing primary concepts, is intended to explore the contribution to termite infestation implied by those essential basic statements.

Case studies are necessarily taken from different countries. It is therefore hoped that this account may be of interest in those lands where heritage is already threatened by termites but where destructive and disruptive treatments to important structures must be minimised, or where resources are not available for sustained programmes of control.

Brian Ridout
June 2022

Acknowledgements

I would like to thank Mathew Green for finding Rentokil archive photographs, and Rentokil Initial plc for permission to use these together with a key to termite families from *Termites in Buildings: Their Biology and Control* (Rentokil Library) Edwards and Mill, 1986 © Rentokil Initial plc.

Plates 1, 4, 7 and 8 have been kindly supplied by Lyle Buss of the University of Florida.

I wish to thank Doug Wechsler for Plate 3 and Kenji Matsuura of Kyoto University for permission to use the caste differentiation figure used here as Fig 2.1.

Iain McCaig of Historic England has generously supplied Figs 3.1, 4.1, 4.3, 4.4 and 7.7, together with much helpful advice.

My special thanks are due to Princess Esther Selassie Antohin and David Michelmore for their introductions to some Palaces and Royal Mausoleums.

Finally, thanks are due to my wife, Elizabeth, and daughter Kate for drawings, comments and general encouragement.

Termites, climate change and the UK – an introduction

Although we are accustomed to think of termites as pests, they are vital for the decomposition of dead organic material and the beneficial modification of soil in tropical and subtropical regions. Approximately 2,600 species of termite have been described, but fewer than 200 species impinge on human activities sufficiently to be considered pests. Of these, only a few cause significant damage to structures and buildings. But though few in number as species, they may be vast in number as individuals and cause considerable damage around the world.

To a termite, the palace of a long-departed king is just another heap of dead wood. People construct and termites destruct – that is the way of things, and retaining a wooden historic building in a land where there are plenty of termites is a battle against nature. This problem is much more complex and severe than the difficulties hitherto encountered by timber conservationists accustomed to beetles and fungi in the UK.

In England, termites have occasionally been found as accidental introductions since at least the 19th century. In 1874, for example, a colony of *Kalotermes* was discovered in a trunk of copal wood (*Trachylobium sp.*) that was imported from East Africa to Kew Gardens Museum (McLachlan 1874).

Gay (1967) compiled a world list of introduced termites, which provides a useful indication of accidental importations into the UK over the 20-year period it covers. Unfortunately, there is no contemporary list. Nevertheless, the old one demonstrates the diversity of termite genera that might be imported. But it is important to differentiate between those which have been brought into an environment they cannot tolerate and ones that might survive to form a colony (*see* Chapter 5).

- *Nasutitermes costalis* (Termitidae) nest found in the cellar of a public house at Spalding, Lincolnshire. This was in packing material from Martinique (Riley 1943).
- *Trinervitermes sp.* (Termitidae) living under a hedge in Croydon, Surrey. Origin unknown (Sweeney 1948).
- *Neotermes jouteli* (Kalotermitidae) from imported *Lignum vitae* from San Domingo in 1955 (Harris 1955).
- *Kalotermes sp.* (Kalotermitidae) tropical species colony found in 1955 in a Bethnal Green timber merchant's yard, London (Harris 1962).
- *Zootermopsis angusticollis* (Archotermopsidae) in Douglas fir imported from Canada (Harris 1956).
- *Cryptotermes brevis* (Kalotermitidae), in balsa wood (*Ochroma pyramidale*) from South America, Lancing, Sussex, 1957 (Gay 1967).
- *Cryptotermes sp.* (Kalotermitidae) in parana pine (*Araucaria angustifolia*) imported from South America, Melton

Mowbray, Leicestershire, 1957 (Gay 1967).
- *Cryptotermes brevis* (Kalotermitidae) in a portable gramophone case from Trinidad, Watford, Hertfordshire, 1960 (Hickin 1961).
- *Kalotermes flavicollis* (Kalotermitidae), Liverpool docks, in wooden boxes containing currants from Greece (Harris 1962).

These records suggest fairly regular accidental importation, and it is likely that many other small infestations would not have been noticed or reported. Six of the imports listed belonged to the family Kalotermitidae. These are known as 'drywood' termites because they live in small, contained colonies and some genera have no need for contact with groundwater.

Drywood termites may live in articles as small as a book and can be difficult to detect. Therefore, they are easily conveyed around the world. *Cryptotermes* species, in particular, cause considerable damage to buildings and their contents throughout the tropics. *Cryptotermes brevis* is now established in Spain, Portugal and the Canary Islands (Nunes *et al* 2009), which are all popular holiday destinations and countries for second homes so that more accidental importation, perhaps in furniture, must be anticipated.

Drywood termite colonies did not thrive and spread because the UK climate was too cool – but times are changing. The top ten warmest years in the UK since 1884 have occurred since 2002 and average temperatures continue to rise (Kendon *et al* 2020).

A study of the effects of climate change on termite distribution (Buczkowski and Bertelsmeier 2017) provided a global risk assessment for 13 of the world's most invasive termites by modelling distribution patterns based on current environmental requirements and predicted climate change. Their distribution maps suggest that the UK climate is already suitable for the drywood termites *Cryptotermes brevis* and *Incisitermes* spp. and becoming more suitable for *Cryptotermes domesticus*. A small colony of the latter species was found in the reptile house of a zoo in the Midlands in 2017. It was living in timber purchased from a local timber merchant 18 months earlier.

In contrast with drywood termites, subterranean termites live in far larger colonies and actively forage over a wide area. Most have to remain in contact with groundwater, although the Rhinotermitidae can form colonies above ground if there is a consistent moisture supply – perhaps from an unresolved maintenance fault. They are more difficult to accidentally import but this does happen. *Reticulitermes flavipes*, a destructive species with a range extending down the east coast of the USA from Canada to Mexico, was first described in 1837 from hothouses at the Royal Palace at Schoenbrunn near Vienna (Gay 1967). A century later it was found infesting the municipal heating system in Hamburg and it is now found over a wide area and is still resisting attempts at eradication.

In 1994 a colony of *Reticulitermes grassei* was found behind a skirting board in the coastal town of Saunton, North Devon, during the installation of a damp-proof course. An initial attempt at eradication using a permethrin-based insecticide was unsuccessful and in May 1998 a thorough investigation was undertaken (Verkerk and Bravery 2004).

This colony contained thousands of individuals and anecdotal evidence suggested that it had been developing for about 25 years. The termites had caused extensive damage to floors, wall frames and roofs in two adjacent bungalows and was subsequently found to extend in the surrounding soil over an area of at least 75m × 35m, encompassing a greenhouse and outbuildings.

The insects were thought to have been brought into the country from Europe in plants or packaging. A thorough biosystematic study of the genus in Europe (Clément *et al* 2001) indicated that the Devon termites were probably imported from south-west France. The closely related *Reticulitermes santonensis* has been recorded in Paris since 1945, and has since spread over much of the city in buildings and living trees (Rasib and Wright 2018). It is in such urban areas that termites have found favourable conditions for survival in the more northern latitudes. In a laboratory study these authors found that the minimum number of worker termites required to establish a successful colony was 250, although survival was possible under stable conditions for as few as 10. Worker termites are only a few millimetres long and 250 is not a large number to accidentally transport.

The Devon colony demonstrated that the subterranean termite genus *Reticulitermes* could survive in the soils of south-west England. However, the infestation was localised and there was little risk of re-infestation from elsewhere, so eradication rather than management was deemed to be a possibility. The government allocated £190,000 to a 12-year eradication programme using 'bait stations' and the growth regulator hexaflumuron (*see* Chapter 7). The termite population fell rapidly in 1999 and no further termite activity was recorded until 2009 (Verkerk and Bravery 2010; *see* Case Study 6 in Chapter 7). It is now thought to have been eradicated, but it is 27 years later!

Our response to an infestation will be to try and destroy it, but £190,000 eradication programmes with baits are not repeatable. Eradication is certainly sensible, but in the process, we would hope to avoid causing more damage than the termites. Fortunately, a study of historic buildings in other lands where termite damage is frequent provides some points worth remembering.

The people who built important buildings generally wanted them to last for as long as possible. They understood the local problems and frequently used materials and construction methods that would maximise longevity. If a building lasted for perhaps a hundred years or more despite wood-destroying fungi and insects, including the local termite population, then the original intentions of the designer were right. If these organisms are starting to invade the building, then something is now wrong and the problem needs to be understood before damaging, costly and possibly unnecessary interventions are embarked upon.

The environment may have changed, perhaps becoming wetter or warmer. The materials available for repairs may also have changed. For example, if we have a building constructed from teak cut from old wild-grown trees, then it will be termite resistant unless modified by fungus. If we replace components with modern plantation-grown teak that is only a few years old, then we are replacing with a material that is full of sapwood and the termites will have little difficulty in destroying it. We may think that we are replacing teak with teak, but we will fail because the problem is more complex. Unfortunately, good-quality termite-resistant timbers are now often difficult to obtain and probably have a high export value so that they are no longer available for local repairs. This makes the conservation of existing timbers even more important.

This is a book about understanding termites, wood and their interactions so that the most appropriate response to a problem can be formulated. Termites can be formidable opponents of conservation, and sometimes, in lands where termites are common, a population has become so established that limiting treatments produces disappointing results. In many other situations the building can be made to perform as originally intended and more of the historic fabric can be saved for future generations to enjoy with limited intervention. If a problem is fully understood, then the most appropriate, economic and potentially successful conservation plan can be devised.

References

Buczkowski, G and Bertelsmeier, C 2017 'Invasive termites in a changing climate: A global perspective'. *Ecol Evol* **7** (3), 974–85

Clément, J-L, Bagnères, A-G, Uva, P, Wilfert, L, Quintana, A, Reinhard, J and Dronnet, S 2001 'Biosystematics of *Reticulitermes* termites in Europe: Morphological, chemical and molecular data'. *Insectes Soc* **48**, 202–15

Gay, F J 1967 *A World Review of Introduced Species of Termites*. Bulletin 286, CSIRO, Australia

Harris, W V 1955 'American termite in imported timber'. *Wood* **20**, 366–7

Harris, W V 1956 'Termites destructive to timber'. *BWPA Convention Record*, 145–77

Harris, W V 1962 'Termites in Europe'. *New Sci* **13**, 614–17

Hickin, N E 1961 'Isoptera in England, a record of accidental importation'. *Timber Technol* **69**, 26–7

Kendon, M, McCarthy, M, Jevrejeva, S, Mathews, A, Sparks, T and Garforth, J 2020 'State of the UK climate 2019'. *Int J Climatol* **40** (51), 1–69

McLachlan, R 1874 'A brood of white ants at Kew'. *Entomol Mon Mag* **11**, 15–16

Nunes, L, Gaju, M, Krecek, J, Molero, R, Teresa Ferreira, M and de Roca, C B 2009 'First records of urban invasive *Cryptotermes brevis* (Isoptera: Kalotermitidae) in continental Spain and Portugal'. *J Appl Entomol* **134**, 637–40

Rasib, K Z and Wright, D J 2018 'Comparative efficacy of three bait toxicants against the subterranean termite *Reticulitermes santonensis* (Isoptera/Blattoidea: Rhinotermitidae)'. *Biomed J Sci & Tech Res* **11** (3), https://doi.org/10.26717/BJSTR.2018.11.002107

Riley, N D 1943 'An occurrence of *Nasutitermes costalis* Holmgren in England (ISOPTERA)'. *Proc R Ent Soc Lond Series A* **18**, 95

Sweeney, R C 1948 'Soldiers of *Nasutitermes* (*Trinervitermes*) (ISOPTERA) in England with a note on artificial termitaria'. *Entomol Mon Mag* **84**, 164–6

Verkerk, R H and Bravery, A F 2004 'A case study from the UK of possible successful eradication of *Reticulitermes grassei*'. Final Workshop COST Action E22 'Environmental Optimisation of Wood Protection', Lisbon, Portugal, 22–23 March 2004

Verkerk, R H and Bravery, A F 2010 *Termite eradication programme – Information leaflet No 19*. UK Department for Communities and Local Government

2 Understanding termites

Termites are remarkable insects. They exist in several forms (polymorphic) and construct nests (termitaria) in which they maintain the optimum environment for the well-being of the colony. The various forms of termite – workers, soldiers and reproductives – are known as castes. Each caste has a different task in a society that communicates and is regulated by interactions that include grooming and 'trophallaxis'. The latter is basically the distribution of nutrients between individuals, but also includes the sharing of pheromones – the passing around of secreted or excreted hormone-like chemicals that produce a response in the colony. Responses may be fairly simple, such as alarm signals and food trails, or complex such as caste differentiation if more soldiers or reproductives are required. A healthy colony thus acts as an efficient self-regulating society.

Termites are eusocial. This means that they have cooperative brood care, overlapping generations and a division of labour into reproductive and non-reproductive groups or castes with appropriate tasks. However, unlike other eusocial insects, an individual cannot live away from the colony. Workers can forage and feed soldiers and reproductives, but cannot reproduce. Soldiers can defend, and reproductives can produce eggs, but neither can feed themselves. This all means that the members of the colony are interdependent, and a termite colony, with its nest, is sometimes considered to be an individual 'super organism'.

2.1 What is a termite?

The termites belong to the order Isoptera. and their development is an incomplete metamorphosis (hemimetabolous). This means that newly hatched individuals, called larvae, are smaller helpless versions of the adult workers (see Fig 2.1 and Plate 3) – there is no pupal stage leading to an adult with an entirely different appearance as found, for example, in butterflies and wasps. They are a sister group to the wood-feeding cockroach genus Cryptocercus, probably diverging from a common ancestor that lived during the late Jurassic period about 170 million years ago. So, termites are cockroaches, but there is ongoing debate about how their complex social organisation developed.

2.2 Classification

Termite families are grouped as 'lower' or 'higher' and as 'dampwood', 'drywood', 'arboreal' or 'subterranean'. These groups are described in the following sections.

The Mastotermitidae, represented by only one living species in northern Australia, is the closest to the original cockroach in body shape, wing venation and the production of egg cases (oothecae). Termite families become more specialised as they progress down the following classification (Engel 2011) towards the Termitidae. The Termitidae include about 75% of all known termite species.

The Engel classification will provide a framework in which to slot the termite genera that are discussed in the following chapters and those that cause most of the damage to timber in heritage buildings throughout the world.

2.2.1 Lower termites

In these termites, cellulose is digested by microorganisms (flagellate protozoa = parabasalids and oxymonads) together with their symbiotic bacteria, in the termite gut.

Family: Mastotermitidae = Subterranean termites
These termites make a subterranean nest around a piece of wood with foraging galleries to find additional food resources. They cause significant damage to buildings in parts of Australia and New Guinea.
Heritage pest genera: *Mastotermes*
Pest status: Significant

Family: Archotermopsidae = Dampwood termites
Dampwood termites live in small colonies in a single piece of rotting wood. Some species may cause damage to timber ends embedded in the ground. They may also invade if there are building or plumbing faults that allow water into the structure.
Heritage pest genera: *Zootermopsis*
Pest status: Minor

Family: Stolotermitidae = Dampwood termites
These termites are closely related to the Archotermopsidae and also live in dead wood.

Subfamily: Porotermitinae
Heritage pest genera: *Porotermes*
Pest status: One species (*Porotermes adamsoni*) is a significant pest of *Eucalyptus* timbers including posts and supports in Australia.

Subfamily: Stolotermitinae
Heritage pest genera: *Stolotermes*
Pest status: Minor – they may invade building timbers if these are wet and starting to decay.

Family: Kalotermitidae = Drywood termites and dampwood termites
Drywood termites colonise dead branches and heartwood in living trees and some invade building timbers. Several genera are a major problem for heritage conservation. Unlike subterranean termites, they do not need to maintain contact with the soil, although some require a damper

environment than others. They live as a small colony, often in a single piece of wood, and colonies of the same species may join together if they meet.

Heritage pest genera (drywood): *Kalotermes*, *Cryptotermes*, *Incisitermes* and *Marginitermes*; **(dampwood)**: *Neotermes*, *Glyptotermes*

Pest status: *Kalotermes*, *Neotermes* and *Glyptotermes* are of minor importance. The other three genera are significant pests.

Family: Hodotermitidae = Subterranean (harvester) termites

These, unlike other termites, are day and night foragers with pigmented bodies and compound eyes. They are specialist grass feeders that remove straw binder from mud bricks, destroy thatch and attack softer timbers.

Heritage pest genera: *Anacanthotermes*, *Hodotermes*

Pest status: Significant

Family: Stylotermitidae = Arboreal termites

These feed on the dead parts of living trees, where they nest in galleries in larger branches and the stem. They are considered to be intermediate between the Kalotermitidae and the Rhinotermitidae.

Heritage pest genera: None

Family: Rhinotermitidae = Subterranean termites

These are found in most environments from the temperate parts of Europe and North America (*Reticulitermes*) to subtropical/tropical (*Coptotermes*) and deserts (*Psammotermes*). All live in subterranean nests associated with wood, and forage over considerable distances.

Subfamily: Coptotermitinae
Heritage pest genera: *Coptotermes*
Pest status: Highly significant

Subfamily: Heterotermitinae
Heritage pest genera: *Heterotermes*, *Reticulitermes*
Pest status: Highly significant

Subfamily: Prorhinotermitinae
Heritage pest genera: *Prorhinotermes*
Pest status: Minor. One species (*Prorhinotermes simplex*) is a minor pest in Florida and parts of the Caribbean.

Subfamily: Psammotermitinae
Heritage pest genera: *Psammotermes*
Pest status: Significant

Subfamily: Termitogetoninae
Heritage pest genera: None

Subfamily: Rhinotermitinae
Heritage pest genera: *Schedorhinotermes*
Pest status: Significant

Family: Serritermitidae = Subterranean termites
This small family contains two genera. One (*Serritermes*) feeds on nest material in sealed galleries within the nest walls of other termites (these are called 'inquilines'), while the other (*Glossotermes*) lives in dry rotten wood.
Heritage pest genera: None

2.2.2 Higher termites
In these termites, cellulose is digested by bacteria in the termite gut.

Family: Termitidae = Subterranean termites
This family contains species that feed on the complete range of plant material from living material to fragments in soil.

Subfamily: Apicotermitinae
African and neotropical rainforest soil feeders with subterranean nests. Some genera have soldiers while others do not.
Heritage pest genera: None

Subfamily: Foraminitermitinae
African soil feeders.
Heritage pest genera: None

Subfamily: Sphaerotermitinae
Central African wood feeders that construct subterranean nests in which they cultivate symbiotic bacteria on structures called 'combs'. These are any cellulose-based material processed by the termites into a substrate garden for bacteria or fungi.
Heritage pest genera: None

Subfamily: Macrotermitinae
African and Southeast Asian wood feeders that construct complex subterranean or epigeal nests (ones below ground with above-ground mounds) with fungus combs.
Heritage pest genera: *Macrotermes, Odontotermes, Hypotermes*
Pest status: Highly significant

Subfamily: Syntermitinae
Neotropical, feeding on plant materials in varying stages of decomposition. Most produce epigeal nests, while others are entirely subterranean, arboreal or 'inquilines' (ones that live in the nests of other termites).
Heritage pest species: None

Subfamily: Nasutitermitinae
Pantropical termites with no visible mandibles and the front of the head projected into a snout (nasus). Many construct arboreal nests, some are subterranean, and a few construct mounds (particularly in Australia). They mostly feed on lichen or dead leaves and branches.
Heritage pest genera: *Nasutitermes*
Pest status: Significant

Subfamily: Cubitermitinae
These are found in the forests and savannahs of sub-Saharan Africa, where they are mostly soil feeders. Nests may be subterranean or mushroom-shaped epigeal mounds.
Heritage pest genera: None

Subfamily: Termitinae
Pantropical distribution with species feeding on the full range from sound to totally decomposed plant materials.
Heritage pest genera: *Amitermes*, *Globitermes*, *Microcerotermes*
Pest status: Significant

2.3 Castes and caste behaviour

Termites, as eusocial insects, are organised, both morphologically and functionally, into castes, which may be reproductive or supporting.

2.3.1 Primary and secondary (neotenic) reproductives
The primary reproductive pair is called the queen and king. They are the founders of the colony and produce eggs which hatch into helpless and dependent larvae. New queens may initially lay large eggs with a short hatching period because the primary pair don't move out of the brood chamber so their internal fats and proteins must be metabolised as the food resource. First-brood workers are consequently undersized, but they are essential to the survival of the new colony. Eggs are a product of male and female parents throughout the life of the colony.

As the population increases, brood care becomes a cooperative undertaking and does not require further input from the queen. This strategy increases reproductive potential because once the first larvae develop, the queen can become a full-time egg producer. Constant attention from the growing workforce means that egg size may decline as egg numbers increase because larvae can be less well developed when they hatch (Matsuura and Kobayashi 2010).

Termite queens are physogastric. This means that their abdomen may swell with eggs to many times its original size. Swelling is accommodated by growth of the soft flexible cuticle (arthrodial membrane) that connects the hard (tanned) abdominal skeletal plates (Bordereau 1982) and this eventually gives the abdomen of the queen a striped appearance (Fig 2.1; Plate 1). Queens of smaller-colony termites such as *Kalotermes* may be little different in size to their mate, but at the other end of the scale, those of *Coptotermes* may reach 70mm in length (Plate 2). Movement for the queen may become almost impossible, but the king's shape does not change.

Full growth may take many years and queens of the Termitidae lay many thousands of eggs per day, which are immediately removed by the workers to nurseries. However, the number of eggs may also be controlled by the needs of the colony. Skaife (1955), in his 20-year study of the South African black mound termite (*Amitermes hastatus*), found that egg laying was regulated by the workers. The queen was neglected during the winter months and laying ceased, but she received more food

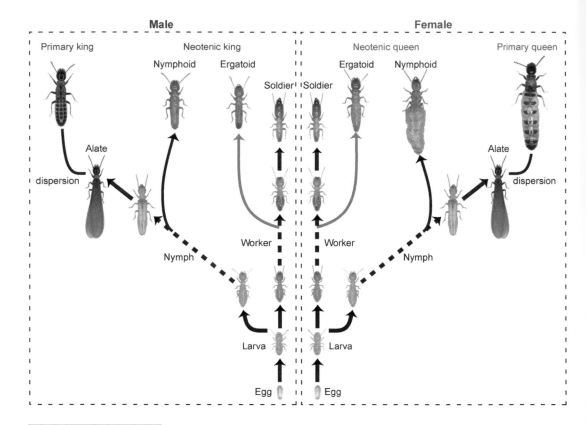

Fig 2.1
Caste differentiation in
Reticulitermes. Eggs hatch
into larvae, which either
develop wing buds and
become nymphs or they
become workers. The
nymphs may develop into
reproductives. The workers
have the potential to develop
into soldiers or secondary
reproductives according to
the needs of the colony.
[Reproduced with permission
from Matsuura et al 2018]

when the warmer weather arrived and egg laying commenced again. If
there was prolonged drought and food was scarce, then egg laying was
halted by withdrawing their attention to the queen. When conditions
became really difficult, the workers used previously laid eggs and young
larvae to feed the colony, thus conserving resources. In actuality, the
control mechanisms are likely to be far more complex than Skaife's early
observations suggest. Nevertheless, he makes the point that the queen
is just the egg-laying department of a complex and carefully regulated
society. The eggs are constantly tended and moved around by the workers
(Plate 3) and the queen has no interest in them. Skaife was unable to
hatch termite eggs away from the attentions of the termites.

Large numbers of eggs have also been reported for the
Rhinotermitidae. Egg production in this family may be enhanced by
supplementary reproductives. If the primary reproductives die or are
unable to maintain the egg requirements of the colony, then nymphs of an
appropriate sex may become mature and replace them. Several substitute
queens may be required to supply the needs of a large colony.

Primary reproductives (but not workers or soldiers) in colonies of
Rhinotermitidae and Termitidae may live for more than 50 years. In other
organisms it is generally found that increasing age reduces fecundity,
but these termite queens seem to have overcome this problem and there
is considerable interest in understanding the biological mechanisms
involved as part of general studies on ageing.

For most other termites the number of eggs laid is probably far smaller – between 200 and 300 per year has been suggested for the Kalotermitidae. Ageing reproductives in these drywood termites are not replaced, and a population will decline with egg production so that a colony may die out in 10–15 years.

Eggs are around 1mm long by 0.5mm wide and are laid singly, except for *Mastotermes* where they are laid in batches of 16–24 cemented together in the manner of cockroach egg cases (oothecae) though without an outer covering.

Eggs hatch into larvae and these develop into nymphs with wing buds (that will grow into wings) or workers. These then differentiate according to the needs of the colony. Figure 2.1 illustrates development pathways for *Reticulitermes*.

Some nymphs become winged primary reproductives known as 'alates' (from the Latin for wing) (Fig 2.2). These are the termites that swarm to found new colonies, although some may be retained in the nest, when they are called adultoid reproductives. These increase the egg-laying potential and size of the colony.

If the king or queen dies or the colony fragments, secondary reproductives may be produced by neotony – the acquisition of sexual maturity in otherwise sterile juvenile forms. Figure 2.1 shows that these secondary reproductives, known as 'neotenics', may develop from the nymph (nymphoid) or from the worker caste that has also developed from the larvae (ergatoid).

Secondary reproduction protects the colony if the nest is damaged and allows expansion into supplementary nests ('budding'). The new reproductive pair inherits existing resources and has other individuals around to assist them (unlike the primary pair). In this way the hazards of alate swarming are avoided, but the colony must inbreed and its fitness may decline.

Soldiers occasionally turn into reproductives in the primitive families Archotermopsidae and Stolotermitidae. These soldier-headed secondary reproductives have shorter mandibles than normal soldiers (Masuoka *et al* 2021).

Fig 2.2
An alate of the European drywood termite *Kalotermes flavicollis.*

Alates (*see* Fig 2.2) – winged reproductives with functional compound eyes – vary in size from genera to genera. Those of the large African *Macrotermes* may have a wingspan of 90mm, while *Microtermes* alates may only be 12mm. Alates develop two pairs of sub-equal wings, which, after a dispersal swarm, break off at their bases at a line of weakness called a humeral suture (Fig 2.3). The resulting wingless kings and queens may form new colonies, but the whole process is a feast for birds and other predators and few reproductives will survive. The trigger for this swarming behaviour seems commonly to be colony maturity and colony size. *Kalotermes* alates, for example, do not appear until the colony has been growing for two or three years. The winged individuals congregate in the nest – perhaps in response to growing antagonism from the worker caste – until conditions are right for the swarm to depart (Plate 4).

The external conditions that finally trigger a dispersal flight seem to be variable, but may include the season of the year and the weather

Fig 2.3
Discarded termite wings
that have broken off at the
humeral suture.

conditions. The nest, which is usually sealed, will be breached to allow the alates to emerge, but this must be for as brief a time as possible because of the risk from ants and other predators. The concealed termites within the nest must somehow recognise that external conditions are optimum for dispersal. Furthermore, emergence of the species colonies must be synchronised so that nests produce alates at the same time. If the members of one swarm do not intermingle with the members of other swarms of the same species, then there will be inbreeding (Vargo and Husseneder 2011) because the members of each swarm are all siblings. Outbreeding – mating between individuals from different populations – is essential to maintain species fitness.

When the dispersal flight occurs, the alates will fly generally rather feebly for perhaps a few hundred metres, then drop their wings, usually by twisting their bodies. The average flight distance for *Coptotermes formosanus* has been recorded as 621m in one study, with a maximum of 12.3km. Sometimes turbulence will carry individuals high into the air, but this considerably reduces their chances of finding a mate. The pressure from predation means that the whole process is necessarily brief, and frequently a scattering of discarded wings is the only indication that swarming has occurred (Plate 5).

In some termites the female raises her abdomen after her wings have detached and emits a sex pheromone to attract a male. This may produce a group of males who will follow the queen and each other in a chain that is termed a 'tandem'. However, a new colony will generally be founded by only one pair, and nest construction may be commenced by both sexes or by the female alone. Only a very small percentage of the dispersed reproductives will survive and successfully establish a new nest.

The production of caste members and the growth of wings in some reproductive nymphs is controlled by a complex of pheromones and nutrients that are exchanged in complex patterns of grooming and trophallaxis – the passing of food or tactile and chemical stimulation between individuals.

The colony maintains a dynamic balance of caste members. More soldiers and reproductives may be produced if necessary and some may be killed if there becomes an excess. More workers may be needed in spring and summer and more soldiers have been found just before dispersal flights in *Reticulitermes*. Cannibalism, if numbers are to be reduced, and the eating of surplus eggs or injured or dead individuals, will play an important role in the conservation of nitrogen and other nutrients. It is the colony that is important, not the individuals.

2.3.2 Nymphs, workers and pseudergates

In summary, young termites or immatures are called larvae, or nymphs after the second moult if they develop wing buds. They may grow into workers or turn into soldiers or reproductives depending on the requirements of the colony.

Workers (which may be male or female) are sterile, wingless, mostly blind and usually unpigmented (Fig 2.4), although their abdomens may appear dark because of the ingested food fragments. Unpigmented species live in total darkness, but those that forage freely (mostly grass, leaf litter or lichen feeders) are black or some shade of brown with functional compound eyes. Large and small individuals of the same caste may be present in the same colony and probably perform slightly different roles. The workers' main tasks are to tend the eggs and larvae, construct and repair the nest, forage for food, and feed the soldiers and reproductives.

Individuals that undertake these tasks to a variable extent in the primitive drywood termites (Kalotermitidae) and the dampwood termites (Archotermopsidae) are considered to be false workers because they can still differentiate into other castes, while the future caste was programmed in the larva/nymph stage for the true workers. These false workers are referred to as 'pseudergates' or pseudoworkers. The amount of work required from pseudergates is probably less than that required from workers because the drywood and dampwood termites do not make complex nests (*see* section 2.4).

If a subterranean termite worker finds a suitable food resource, it lays a trail to it by pheromones secreted by sternal glands on its underside. If the food source and water supply are sufficient, a 'foraging tube' may be formed from chewed wood, substrate fragments and dry faecal pellets,

Fig 2.4
Foraging worker termites
(*Reticulitermes balkanensis*)
with transparent abdomens.

all held together with faecal cement. Tubes are generally about 10mm wide and may extend for many metres. Construction is rather similar to bricklaying. The termite deposits a blob of wet faecal cement, then turns around and positions the pellet of tube material with its mandibles. Tubes provide concealment but they also allow an appropriately humid environment to be maintained. The tube material is sometimes spread out into a sheet over the food source (Fig 2.5).

Fig 2.5
Foraging tubes and sheets
of tube material, produced
by *Amitermes messinae* on
a ceiling construction in
Bahrain.

Most termite workers are blind and the use of pheromone trails to a food resource is not necessarily efficient. The tube trail may lead the termites up a tree and down again to reach a food source that was a much shorter direct distance from the nest.

If an intruder or predator is encountered in the nest, then an alarm sequence will be triggered. Alarm signals are mostly produced by both worker and soldier termites, but in some more advanced species they may be restricted to soldiers. They transmit signals by jerking their bodies to and fro or head banging to produce substrate vibrations, bumping between individuals, and laying a pheromone trail. Nest repair behaviour is also stimulated, and if the intruder is immobilised, perhaps by the soldiers biting off its legs, then the workers will bury it in nest material.

Competition between termite colonies can be fierce in some genera, but apart from this, ants are the major invertebrate foe. The termites are soft bodied and much slower moving. Skaife (1955) compared the speed of *Amitermes hastatus* (called by him *A. atlanticus*) with published speeds of an American ant *Liometopum apiculatum* that lives in a similar type of environment. He found that his termites moved at a steady 45cm/minute at 30°C, while the ants achieved 180cm at the same temperature.

A few termite genera that feed on soil do not have soldiers, but the workers are able to protect the colony by exploding their abdomens to release a sticky and probably toxic substance which entangles and kills both the termite and its attackers (Sands 1982). This is a process known as 'autothysis', which blocks tunnels with dead bodies. Worker suicide under intense provocation may also occur when there are soldiers in the colony. Old workers of *Neocapritermes taracua* (Termitinae) develop pouches just below the cuticle of the first abdominal segment on their backs (tergite) that contain a blue copper-based protein. When the termite voluntarily ruptures its internal organs and bursts, these protein crystals react with salivary gland secretions to produce toxic compounds (benzoquinones). This ability increases with age as the worker's foraging capability declines (Bourguignon *et al* 2016).

2.3.3 Soldiers

Workers can feed and provide the day-to-day requirements of the colony, including nest and gallery construction and repair, but their defence capability is frequently directed at removing vulnerable members of the colony to safety. Social organisation requires individuals that can provide aggressive protection, particularly against their ant enemy. This role is normally undertaken by soldier termites (Plate 6). The percentage of soldiers in a colony is variable but commonly around 5%, although in some soil-feeding genera there are none, while in *Coptotermes* (for example) they might exceed 30%. Soldiers are sterile and have a body shape that does not differ much from the workers, but the head structures are variously modified (Fig 2.6) to defend the colony. Heads may be swollen to carry the muscles required for massive mandibles with prominent teeth used for crushing as in Mastotermitidae and Hodotermitidae. Others have mandibles adapted for snapping or slashing, or have smaller mandibles for attaching themselves to an invader. A few have plug-shaped heads (phragmotic) that are thought to be used to block gaps in the workings or nest when necessary (as in *Calcaritermes*

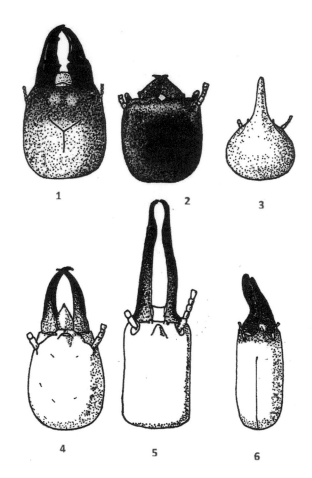

Fig 2.6
Soldier heads showing the
range of adaptations.
1 Crushing jaws of
Zootermopsis anqusticollis
2 Phragmotic head of
Cryptotermes domesticus
3 Nasute head of *Nasutitermes
novarum-hebridarum*
4 Slashing jaws of *Coptotermes
grandiceps*
5 Symmetrical snapping jaws
of *Termes odontomachus*
6 Asymmetrical snapping jaws
of *Pericapritermes dumicola*
[adapted from Harris 1961]

and *Cryptotermes*). Soldiers in the primitive Hodotermitidae and Archotermopsidae families have compound eyes, but most soldier termites are blind.

Reduction in the muscles required to operate heavy crushing mandibles means that defence may also be chemical using glandular secretions delivered via head modifications. Termites of the more advanced families Rhinotermitidae, Serritermitidae and Termitidae have a frontal gland on the top of the head. It opens as a pore (fontanelle) along the midline on top of the head, and is particularly well developed in some soldiers of the Rhinotermitidae (Plate 7). Frontal gland secretions are used for defence and may be contact poisons, repellents and irritants, glues or alarm pheromones. The pore in *Coptotermes*, for example, secretes a white liquid which thickens in the air into glue entangling both the termite and the predator. Skaife (1955), who made a thorough study of *Amitermes hastatus*, found that their secretion had a powerful irritant effect upon the ants and if they were wetted sufficiently by it, they were rapidly killed.

Globitermes sulphureous stores a yellow liquid in a reservoir that fills the frontal part of the abdomen. The soldiers throw this from their mouths when attacked, by violently contracting their abdomens, and the liquid congeals in the air entangling their adversary. The violence of the defence mechanism ruptures the abdominal wall and is suicidal (autothysis, as described previously). However, this is only part of a defence strategy that includes the effective use of mandibles and pheromones to recruit more soldiers (Bordereau *et al* 1997).

In the family Nasutitermitinae, the head is extended into a pear-shaped nozzle with the fontanelle opening at the tip and the mandibles are reduced or absent. The head is used as a reservoir for the frontal gland secretions and the mandibular muscles are used to project the contents with considerable accuracy over a distance of several centimetres. Skaife (1955) investigated defence in *Trinervitermes gemellus*, and his termites shot fine white threads of secretion at ants. These threads were so slender and light that they floated in the air and if they touched an ant, then it rapidly retreated. An extract made by grinding up a few soldier heads on a microscope slide and tested against ants was rapidly fatal. The soldiers are blind and so the accuracy presumably depends on odour and surface vibrations.

Soldiers of the two indigenous European termite genera *Reticulitermes* and *Kalotermes* should be readily differentiated by mandible characteristics, but this requires a microscope. A useful characteristic for field identification is the more cylindrical shape of *Kalotermes* (Plates 8 and 9).

2.4 The nest

Most termites construct their own nests, which may range from simple excavated galleries in wood, large excavated subterranean cavities filled with a complex nest structure, to arboreal nests and mounds. Some termite mounds are immensely complicated constructions that have been a popular topic for study. These nests are built to maintain an environment that is controlled by the colony so that termite development is generally insulated from the outside world.

2.4.1 Nesting behaviour

There are three main categories of nesting behaviour. In the simplest form, chambers and galleries are excavated in wood, which may be above ground (Fig 2.7), by drywood termites (Kalotermitidae) and dampwood termites (Archotermopsidae). A colony will live in a single piece of wood or several pieces in close contact. The reproductives and their colony stay within the wood that they have invaded, using it both as a food source and for shelter. Openings to the exterior are provided solely for the departure of alates and the removal of excess faecal pellets and are sealed again after use. High humidity within the nest is conserved by restricting its size using faecal pellets to seal off galleries that are no longer required. These termites are known as 'one-piece nesters'.

Termites in the second category are called 'intermediate nesters'. They nest in wood and seek out other suitable timbers, which may also be used for nesting when the original has been consumed (Fig 2.8).

Reticulitermes, *Heterotermes* and *Coptotermes* (Rhinotermitidae) are examples of this category. They nest in buried wood and construct a system of galleries, both between colonised pieces of wood and as foraging tubes to find fresh food resources above ground. Nests may

Fig 2.7
Drywood termites (*Kalotermes flavicollis*) have constructed a 'one-piece' nest in the base of a door frame. They do not have to forage because they feed on the wood they are nesting within (Skopelos, Greece).

be filled with a sponge-like material consisting of wood fragments and faeces. This is known as carton. These colonies may grow very large and their dispersal is aided by replacement reproductives so that the colony fragments. *Coptotermes* species may be present in huge colonies of a million or more individuals producing foraging trails that may extend for 150m. *Mastotermes* (Mastotermitidae), and sometimes members of other families, seem to have a similar nesting behaviour. This category contains important destroyers of heritage structures.

The harvester termite (Hodotermitidae) nest is also formed from discrete chambers, sometimes at a considerable distance underground, that are connected by a network of galleries.

Termites in the third category nest separately from their food resource and are known as 'separate nesters'. The family Macrotermitinae contains examples of these. They construct their nests from substrate material mixed with saliva. The nest contains a cell for the queen, nurseries, chambers and galleries, all grouped around a fungus comb (Fig 2.9). Covered foraging tubes above ground are constructed from

Fig 2.8
Macrotermes gilvus termites have formed a mound around a bench, which they are using as a primary food resource. They will forage elsewhere from this base (Malaysia).

Fig 2.9
Excavated 'separate nest' of *Odontotermes sp.* They forage away from the nest to find a wood source for food (Ethiopia).

the nest material, for example sand or clay moistened with saliva, in the genus *Odontotermes.*

A further category (not illustrated) includes termites that live in small colonies in sealed galleries within the nests of other types of termites. These are called 'inquilines' and they feed on the nest material and stored reserves of their unwilling host. They will be attacked by the host termites if found.

2.4.2 Nest types and construction

The lower termites are one-piece nesters or intermediate nesters. The higher termites (Termitidae) produce a wide range of nest types, from subterranean to arboreal and mounds, depending on genera. All above-ground mounds (epigeal nests) are thought to develop from subterranean nests. Some arboreal nesters such as *Nasutitermes*, make nests in trees, but also maintain contact with the ground.

Nests may be formed from materials such as particles of soil or wood transported to the nest (exogenous materials) and moistened with saliva or faecal material or both. Wood-feeding *Amitermes* and *Microcerotermes*, for example, use their excreta to form nests made from carton, which may also contain sand or some other substrate material. The complex nests of higher termites often have a thicker outer part constructed from inorganic materials and inner parts from faecal material. There may be an inner 'royal chamber' constructed to resist mechanical damage. Mounds may have outer galleries that are narrower than the inner so that they are easier to defend. They may also have a complex ventilation system. This is driven by the metabolic heat of the colony, which creates a stack effect drawing air in through openings or creating air exchange by diffusion through the porous nest walls and galleries.

In these days of pandemics, it is particularly interesting to see how other creatures manage overcrowding. The *Coptotermes* nest construction, for example, tends to maintain a constant temperature and high relative humidity, which should have adverse consequences. A dense population nesting in damp ground exposes the termites to pathogenic fungi and the termites have had to evolve ways to counteract this. These include immunological, biochemical and behavioural responses. Some of these are of particular interest. For instance, the nest of *Coptotermes formosanus* is constructed from carton, which is chewed wood particles and faecal material. There is also a faecal lining to the foraging galleries. This faecal lining continues to be metabolised by micro-organisms and has been found to contain Actinobacteria, mostly *Streptomyces*, which suppress the growth of pathogenic moulds (Chouvenc *et al* 2013). The lining is then re-ingested by the termites. (Over two-thirds of our clinically useful antibiotics of natural origin are derived from *Streptomyces*.)

Another investigation found that nest material of *C. formosanus* contained the volatile hydrocarbon naphthalene (Chen *et al* 1998). This is a fumigant which kills, repels or inhibits insects, moulds and parasitic worms. Tests with fire ants (*Solenopsis invicta*) showed that the ants, which are an important predator, were paralysed at concentrations of naphthalene that had no visible effect on the termites. There were also indications that naphthalene was used for trail following.

These factors, together, may be a major reason why attempts at biological control of termites have generally been unsuccessful.

References

Bordereau, C 1982 'Ultrastructure and formation of the physogastric termite queen cuticle'. *Tissue Cell* **14** (2), 371–96

Bordereau, C, Robert, A, Van Tuyen, V and Peppuy, A 1997 'Suicidal defensive behaviour by frontal gland dehiscence in *Globitermes sulphureus* Haviland soldiers (Isoptera)'. *Insectes Soc* **44**, 289–97, https://doi.org/10.1007/s000400050049

Bourguignon, T, Šobotník, J, Brabcová, J, Sillam-Dussès, D, Buček, A, Krasulová, J, Vytisková, B, Vytisková, B, Demianová, Z, Mareš, M, Roisin, Y and Vogel, H 2016 'Molecular mechanism of the two-component suicidal weapon of *Neocapritermes taracua* old workers'. *Mol Biol Evol* **33** (3), 809–19, https://doi.org/10.1093/molbev/msv273

Chen, J, Henderson, G, Grimm, C, Lloyd, S W and Laine, R A 1998 'Termites fumigate their nests with naphthalene'. *Nature* **392**, 558–9, https://doi.org/10.1038/33305

Chouvenc, T, Efstathion, C, Elliott, M and Su, N-Y 2013 'Extended disease resistance emerging from the faecal nest of a subterranean termite'. *Proc R Soc B: Biol Sci* **280** (1770), 20131885, https://doi.org/10.1098/rspb.2013.1885

Engel, M S 2011 'Family-group names for termites (Isoptera)'. *Redux ZooKeys* **148**, 171–84

Harris, W V 1961 *Termites: Their Recognition and Control*. London: Longmans

Masuoka, Y, Nuibe, K, Hayase, N, Oka, T and Maekawa, K 2021 'Reproductive soldier development is controlled by direct physical interactions with reproductive and soldier termites'. *Insects* **12** (1), 76, https://doi.org/10.3390/insects12010076

Matsuura, K and Kobayashi, N 2010 'Termite queens adjust egg size according to colony development'. *Behav Ecol* **21** (5), 1018–23

Matsuura, K, Mizumoto, N, Kobayashi, K, Nozaki, T, Fujita, T, Yashiro, T, Fuchikawa, T, Mitaka, Y and Vargo, E 2018 'A genomic imprinting model of termite caste determination: Not genetic but epigenetic inheritance influences offspring caste fate'. *Am Nat* **191** (6), 677–90, https://doi.org/10.1086/697238

Sands, W A 1982 'Behaviour of African soldierless Apicotermitidae (Isoptera: Termitidae)'. *Sociobiol* **7**, 761–72

Skaife, S H 1955 *Dwellers in Darkness*. London: Longman Green

Vargo, E L and Husseneder, C 2011 'Genetic structure of termite colonies' *in* Bignell, D E, Roisin, Y and Lo, N (eds) *Biology of Termites: A Modern Synthesis*. London: Springer, 321–47

3 Termites and water

3.1 Moisture requirements

Desiccation is always a problem for small soft-bodied insects that have a large surface-to-volume ratio. Their exoskeleton is made from chitin, which is permeable to water, and most of the cuticle of termites is pale and untanned (unsclerotised). Insects use a mixture of sclerotin precursors and lipids in the outer layer of their exoskeleton to limit moisture loss, and these lipids also form a thin external wax waterproof coating (Wigglesworth 1985). Most termites are particularly prone to moisture loss and water resources must be sought and conserved. These resources may be metabolic water produced by the breakdown of their food material or free water obtained directly from a damp food resource or the environment.

3.1.1 Metabolic water

Water and carbon dioxide react together using sunlight as an energy source to produce glucose and oxygen during photosynthesis.

$$6H_2O + 6CO_2 = C_6H_{12}O_6 + 6O_2$$

Cellulose is a straight-chained polymer formed from several hundred to over 10,000 of these glucose molecules strung together.

This all means that cellulose-based materials, even the desiccated plant debris sometimes utilised by desert-living sand termites (*Psammotermes hybostoma*) are potentially a significant reservoir of water if they can be metabolised back into water and carbon dioxide (Turner 2006).

Some drywood termites (Kalotermitidae) seem to be more dependent on metabolic water than other termites. These form small colonies and live within their food resource. There is no contact with the ground. Their gut systems contain six large rectal pads for moisture re-absorption (Zukowski and Su 2020) and produce dry faeces in the form of six-sided pellets (Fig 3.1 and Plate 10). Nevertheless, even the Kalotermitidae form a very mixed group when ancillary moisture requirements are considered, and faeces may range from dry pellets to paste in those genera that favour a damper environment (dampwood termites).

A comparative study in 2020 (Zukowski and Su 2020) found that while the subterranean termite *Coptotermes formosanus* required a very high relative humidity to survive, the drywood termites were tolerant across a wide lower-humidity range (18%RH to 73%RH). At the humid

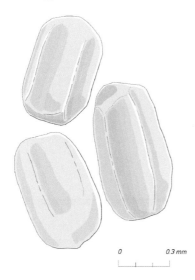

Fig 3.1
Faecal pellets of *Kalotermes flavicollis*. These are dry and have six slightly concave facets caused by the faecal material squeezing between the six large rectal pads that absorb moisture.
[© Iain McCaig]

0   0.3 mm

end of the tolerance scale was the American pest species *Neotermes jouteli*, a 'drywood termite' that is found in damp wood. It still produces faecal pellets, but these tend to be soft and clumped together. These termites could tolerate dry conditions but they steadily lost body weight – probably because of the metabolic breakdown of their fat reserves. This body weight could be regained, and the tolerance would presumably make them resilient to changing conditions.

At the other end of the range was *Cryptotermes brevis*, which usually lives in dry and sound pieces of timber with moisture contents below about 15%, and is therefore easily transported around the world. This species is generally found as a serious pest within buildings. Survival with *C. brevis* was adversely affected by high humidities, a phenomenon that has been termed 'water poisoning' (Woodrow *et al* 2000) which, if maintained, would produce a high equilibrium moisture content in the wood. Experiments, however, have shown that *C. brevis* can slowly acclimatise to high humidities (Steward 1981), but these become more lethal as temperatures increase. If the termites have to cope with high humidities, then they move to the lowest temperature. This is perhaps because the air moisture content is also increasing with the temperature. A study of the distribution of *Cryptotermes* species in Africa (Williams *et al* 1976) concluded that *C. brevis* was found in cooler locations than were tolerated by the other species, while a study of six species of *Cryptotermes* (Steward 1982) indicated that *C. brevis*, unlike the other species which preferred an elevated humidity, was capable of maintaining a high level of egg laying (reproductive potential) across the entire humidity range tested (52%RH–92%RH). This is a termite that may become widespread across Europe.

The European drywood termite (*Kalotermes flavicollis*) requires intermediate conditions. Its natural habitat is the centre of decayed trees, and it generally attacks timber when a source of moisture has allowed decay to become established. Insects that invade buildings tend to do so when their natural environmental requirements are replicated. *Kalotermes* damage is generally found in locations such as the bases of door jambs, windowsills or strip flooring in front of windows, where there is ongoing wetting from weatherproofing faults (Plate 11). The termites also conserve humidity by sealing off galleries that are no longer needed with heaps of pellets.

3.1.2 Environmental water

The ability to resist desiccation seems to vary from genus to genus and perhaps even from species to species. This ability may depend on many factors, including the thickness of the cuticle, the mix of waterproofing lipids and the size of rectal pads for re-absorption in the gut. But generally, the dampwood termites and the subterranean termites require a high humidity. A study of *Coptotermes formosanus* and *Reticulitermes flavipes*' survival at combinations of temperature and relative humidity (Wiltz 2015) found that a combination of both environmental factors was essential. Survival was maximised at 10°C and 99% relative humidity (the next humidity category in the experiment was 85%), and a very high humidity became progressively more important as the temperature increased. Mortality occurred rapidly at 65%RH and was not much improved at 75%RH.

A requirement for a high relative humidity in these termites does not seem surprising because they use an excreted semi-liquid faecal paste or saliva in their constructions, and activities generally decline when humidities drop.

Subterranean termites, therefore, tend to nest in places that provide a cooler and more humid environment. This may be because the nest location is damp, or if there is a mound then the construction regulates the environment. Turner (2001), working with *Macrotermes* in the arid savannah of South Africa, found that there was always a negative humidity gradient from the mound to the outside, irrespective of time of the day or season. Ventilation via the mound, which was essential for regulating gases to maintain the fungus combs in the nest below, caused evaporation. The moisture lost had to be replenished by water wicking into the nest from the surrounding soil and by moisture transportation by the termites. He calculated that wicking and transportation were essential because metabolic water could only contribute a maximum of 10% of the moisture requirements (Turner 2006).

Information on the depths to which termites might delve in the ground to transport moist materials and water has been obtained from the analysis of soils in termite mounds. These investigations can provide important prospecting indicators for metals in the soil beneath the mound. A study in Niger found that the method was particularly useful when gold-bearing strata were deeper than 2m (Gleeson and Poulin 1989). These authors quote observations from the Zambian copper belt where material known to come from horizons 3m–46m deep was brought to the surface by termites. West (1970) reported that an exploration shaft, dug following the finding of gold traces in a mound, intercepted the termites swollen with water at a depth of 24m. Full termites were proceeding towards the surface, while empty termites were heading down to a water-bearing fissure. It is evident that the termites are capable of burrowing down many metres if the substrate allows.

The majority of termites, even including many drywood termites, would seem to need an ancillary supply of moisture. Termites that produce mounds (epigeal) start with a subterranean nest and then move upwards as colony size increases. The shape of the mound will vary according to local environmental conditions (Korb 2011), but the purpose remains to protect the colony and to provide a buffer against environmental fluctuations by thermo-regulation, gas exchange (to remove accumulating carbon dioxide) and moisture conservation. Mounds may also allow the storage of food reserves. Termite nests, which vary considerably in complexity, are not static, and sections may be constructed or demolished according to the need for homeostatic environmental control.

Arboreal termites nest in trees, but maintain contact with the forest floor and its damp rotting wood resources via forage tubes that run down the tree trunk and spread out on or under the ground. The tubes provide shelter from predators such as ants and protection from the desiccating effects of air movement while maintaining a humid environment as damp food particles are conveyed up the tree (Fig 3.2).

The Rhinotermitidae usually build diffuse subterranean nests around tree roots, buried logs or damp timber, from which they may forage up into buildings using tubes (Plate 12) for protection and to maintain

Fig 3.2
Arboreal termite nest of
Nasutitermes sp. with a
network of foraging tubes
(Malaysia).

humidity. Small nests may be connected by galleries or tunnels and numerous supplementary reproductives are produced so that colony fragmentation is easy, allowing these termites to cause serious damage to historic buildings. The termites will form a nest within the building construction if there is a suitable continuous source of moisture available. These ancillary subterranean termite colonies above ground should be fairly easily destroyed if the sustaining water source can be identified and removed (see Case Study 1).

3.2 Accidental sources of moisture

3.2.1 Flat roofs and roof washing
Pitched roofs are an advantage in countries where there is snowfall or frequent heavy rain, but flat roofs are normal in Middle Eastern and other countries where thermal performance is the primary concern and drainage is of secondary importance. These flat roofs, particularly if the surface area is enlarged by curved surfaces, dissipate heat by thermal radiation.

Water from rainfall, or perhaps from excessive surface washing to settle dust, will evaporate because of solar energy, but moisture loss is a two-way process. Solar heat raises the temperature. Once the surface dries out, capillary attraction can no longer operate and direct evaporation ceases. As the roof warms, the vapour pressure of the moisture that has penetrated beneath the surface will increase and moisture will follow pressure gradients in accordance with the second law of thermodynamics – from high to low. This can take it outside or drive it in, particularly if the building is air conditioned (Fig 3.3). A similar phenomenon also occurs in cooler-climate buildings, where it is called 'warm weather condensation'.

If a roof is frequently washed, then moisture can accumulate on the underside of the slab, producing conditions that can be favourable for termites (Fig 3.4). The problem can be exacerbated by attempts at waterproofing if the added layer is poorly designed (Case Study 1) or maintained (Fig 3.5) so that moisture becomes trapped and vapour movement can only be downwards.

Fig 3.3
Average monthly external vapour pressure (kPa) for Muharraq, Bahrain (see Case Study 2) calculated from published data (climatemps.com). Dotted line shows vapour pressure produced in an air-conditioned internal environment if it was maintained at 24°C and 50%RH. Moisture moves from a higher vapour pressure to a lower, drawing it into the building during the summer months.

Fig 3.4
Underside of a flat roof in Sharjah, United Arab Emirates. Boards are damp-stained and large sections have been destroyed by termites because of moisture migration down through the roof slab.

Fig 3.5
The flat roof of a museum in Iraq. A leaking pipe joint has allowed water to accumulate around open joints between roof lining slabs. This would penetrate and become trapped.

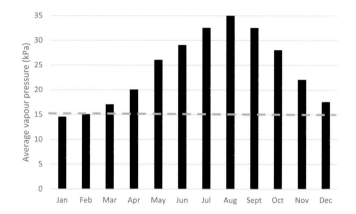

Case Study 1: Museum in Tangier, Morocco

History

This building, which is in the Medina in Tangier, was constructed in 1821.

In 2002 termites (*Reticulitermes lucifugus*) were found to be infesting cartons of books in a room that had been constructed by partitioning a section of the roof on the third floor (Fig 3.6). A survey in 2005 found that the termites had formed a colony in a ceiling beam (Figs 3.7 and 3.8) adjacent to an internal rainwater pipe, in the same area but on the floor below. The termites had swarmed in this room during the previous February. Termite damage continued down the building in bookshelves and door frames.

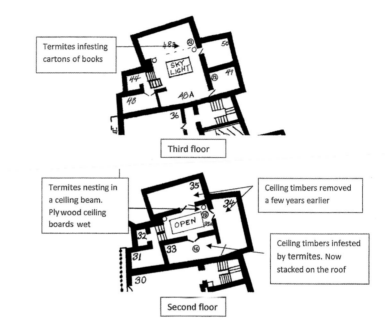

Termites infesting cartons of books

Third floor

Termites nesting in a ceiling beam. Plywood ceiling boards wet

Ceiling timbers removed a few years earlier

Ceiling timbers infested by termites. Now stacked on the roof

Second floor

Fig 3.6
The location of termites on the upper floors of a museum in Tangier.

Fig 3.7
Active termite infestation in an embedded beam end, located in Figure 3.6.

Fig 3.8
Active colony of *Reticulitermes lucifugus* living in the beam end shown in Figure 3.7. The granular material lining the galleries has been transported from the roof slab.

Construction and wood

We were informed that the original timbers would have been cedar (*Cedrus atlantica*) which the Timber Research and Development Association (TRADA) classes as durable. The occupants at the time of our survey took over the building in 1976 when it had been empty for a few years and much of the timber had to be replaced.

The termite-infested ceiling timber (*see* Fig 3.7) was sectioned and found to be maritime pine (*Pinus pinaster*). The heartwood of this species has only a moderate resistance to termites and the sample had a very wide early wood which termites preferentially colonise (1mm late wood/8mm early wood). TRADA classes the heartwood as slightly durable.

An integrated approach to management must take account of timber availability for repairs and so samples were obtained from the local wood yard. The most plentiful and inexpensive softwood was 'pino rocho'. A sample of this was sectioned and found to be European redwood, also called Scots pine (*Pinus sylvestris*). This timber would not have much natural resistance to termites.

Cedar was still available, but was more expensive.

Source of moisture

The tiles on the flat roof had been recently laid because damp in the ceiling boards below was thought to be caused by water soaking through the roof construction (*see* section 3.2.1). The remaining problem seemed to be that the rainwater outlet, which drained half of the flat roof and two adjacent pitched roofs, was too narrow (Fig 3.9) and the internal rainwater pipe it led to did not penetrate through the roof slab to the outlet. This meant that the section through the roof slab was unlined. Water entering the outlet could therefore penetrate sideways into the roof slab, particularly if the system was overloaded and became partially blocked. The roof surface was also not particularly level, and we were informed that a significant depth of water could accumulate when there was heavy rain. This water was then trapped in the roof slab by the tile layer and could not evaporate to the exterior.

The slab around the outlet remained damp enough to allow the termites to flourish despite the lack of rain at the time of the survey.

Fig 3.9
Sloped roofs drain to a tiled slab around a roof light. The drainage outlet through the slab is small and unlined.

Pest management

Termites on the upper floors were a localised problem that had been caused by water penetration from roof drainage and the rainwater disposal system. There were no indications that the termites could infest where conditions remained dry, and the building was well maintained so that most of the timbers would be too dry for subterranean termites. These termites are usually associated with ground moisture, and it was only water penetration into and retention in the roof slab that had allowed them to flourish far above ground level.

There was no advantage in chemical treatment, or the loss of undamaged ceiling timber, provided that the water penetration could be halted.

3.2.2 Air conditioning

Vapour pressure in building spaces is generally very similar to or a little higher than the vapour pressure outside the building. Air conditioning, however, can enhance the vapour pressure gradient so that moisture tends to move towards the interior as shown in Fig 3.3.

Air conditioning units can also contribute considerable amounts of water to the bases of walls via the production of condensate which is generally drained via a pipe through the wall (Case Study 2). This can make the ground and the base of the wall moist enough to sustain termite activity.

Case Study 2: Muharraq, Bahrain

History

Concern over the continuing loss of traditional buildings through rapid urbanisation promoted a conservation study, which commenced during the 1990s. One of the problems identified was that termite damage appeared to be increasingly common. The insects were living within the walls of the houses and forming foraging tubes that disfigured the wall surfaces.

Fig 3.10
Section of original wall showing rubble construction and bonding timbers. The lowest bonding timber is nearly at ground level.

Fig 3.11
The end of a date palm bonding timber. Date palm wood has no resistance to termites.

Construction and wood

Yarwood (1999) states that walls of traditional houses were constructed from coral blocks that were cut from the sea at low tide with picks and brought to site as unshaped rubble. This material was apparently very salty, which must make it hygroscopic at elevated

humidities. Palm trunks were quartered and used as wall plates and embedded tie beams at a vertical spacing of about 0.7m. Palm wood is highly susceptible to termite damage. Ceiling joists were mangrove wood, which has a high resistance to termites, and this will be discussed in Chapter 4. Figures 3.10 and 3.11 show typical construction and embedded timber.

Termites

Identification of the termites involved can provide a basic understanding of a problem that may assist in the formulation of a control strategy.

The majority of the infestations in the traditional houses inspected at Muharraq were found to be two species of higher termites (Termitidae), *Amitermes messinae* and *Amitermes stephensoni*. Harris (1961) states that *Amitermes* is a mound-building genus with an ability to survive in arid conditions. He notes that the nest is generally a cellular construction of carton coated in a layer of soil. A mud layer over a mound will help the retention of moisture.

A revision of the genus from Africa and the Middle East (Sands 1992) notes that *A. messinae* produces dark carton nests that are frequently found to incorporate remains of dead stumps. Hocking (1965) had previously found *A. messinae* in the earth-filled stumps of acacia and myrrh (*Commiphora sp.*) in Tanzania. The species was also found in cattle dung covered with earth.

The nest of *A. stephensoni* was unknown at the time of Sands' revision, and the specimens he studied had been collected 'in tunnels associated with woody plant material'. Badawi *et al* (1986) found the species, together with *A. messinae* and *A. vilis*, living in dead tamarix and date palm in Saudi Arabia, while Al-Zadjali *et al* (2006) recorded it as a pest of date palm in Oman.

Thick walls containing date palm timbers would seem to be an appropriate habitat for the *Amitermes* termites. If the blocks were contaminated with salt, then this would keep the environment within the wall moist at high humidities, and investigations have shown that some termites are salt tolerant (Chiu *et al* 2021).

Source of water

The most likely cause for an increase in termite activity would be a change in building usage and the most relevant seems to be air-conditioning condensate (Fig 3.12). This water will soak into the base of the wall where there may be bonding timbers (*see* Fig 3.10), and some moisture will be drawn inwards by the vapour-pressure gradient. It will provide an easily accessible resource for foraging termite colonies.

Fig 3.12
Condensate from air-conditioning units forming puddles at the base of a wall in Muharraq during early September.

Pest management

The first option to try would be to collect and perhaps utilise the condensate from the units. A foam-formulated termiticide treatment may then be appropriate (*see* Chapter 7), possibly targeted by using a thermal imaging camera to locate timbers and perhaps termite activity (*see* section 6.2.4).

3.3 Construction failure

An important building might be constructed from the best available materials, and drainage may be devised to take the water safely away from the building, but continued resistance to termites depends on the ongoing functioning of the drainage. If this fails, then termites will invade the timbers.

Neglected maintenance implies criticism in wealthy western countries, but may be inevitable in lands where resources are very limited. Warfare and political turmoil can be very destructive, or perhaps old buildings are viewed as a legacy from times best forgotten. Religious buildings of various kinds have also tended to be neglected because the function of the building was more important than the preservation of the structure. For example, we made an infestation survey of a large skete on Mount Athos (Greece). The survey was not satisfactorily completed because there was never anyone available to facilitate access. The monks slept for eight hours (in aggregate), prayed for eight hours and worked in the gardens for eight hours – when a ceiling collapsed, they moved to a different room.

Now, however, as most countries are accessible to travellers, this historic heritage may be seen as an asset if tourism becomes an important part of a country's economy. The resources that are available have to be used carefully and an assessment of how the drainage was supposed to function is needed so that effective and sustainable repairs can be undertaken. An example is provided in Case Study 3.

Case study 3: The palace of King Kumsa Moroda, Negamte, Ethiopia

History

This building dates from about 1870 and was occupied until a communist dictatorship forced the royal family to depart 100 years later. The building is now in a somewhat dilapidated condition with significant termite damage. Comprehensive termite treatment would be difficult and probably not effective for long because the termites causing the damage are very plentiful in the area. However, the building was constructed as a permanent structure that would resist termites and it was still largely intact after nearly 150 years, which included a long period of unavoidable neglect. Corrugated-iron-roofed buildings in Ethiopia generally last for less than eight years (Abdurahman 1990). This longevity of the palace suggests that a return to the way in which the drainage originally functioned would greatly assist in stabilising the situation.

Construction and wood

The main house and a series of round pavilions were constructed on a rubblestone plinth that raises them several metres above the ground (Figs 3.13 and 3.14). The earthen walls of the buildings are protected from the weather by the wide overhang of the corrugated iron roofs, which also cover wide timber balconies on all sides.

Fig 3.13
The main house and pavilions are constructed on a rubblestone plinth so that they are several metres above the ground. The earthen walls are protected by a wide roof, which oversails the wide balconies.

Fig 3.14
The pavilions are connected by terraces that drain through the plinth wall.

The type of timber used in the construction is unknown, but its survival is evidence that it should be termite resistant. Microscopic examination of samples by the Royal Botanic Gardens Herbarium at Kew (UK) could only suggest that the wood was some genus of the family Sapotaceae.

Termites

The termites that we found attacking the structure were a species of *Odontotermes*. A survey of termite-damaged housing in the central region of Ethiopia found that most of the problem was caused by *Odontotermes* and the closely related genus *Macrotermes* (Debelo and Degaga 2014). Both of these belong to the Termitidae and are large-colony-forming subterranean termites that construct fungus combs. *Macrotermes* constructs a mound, but the *Odontotermes* nest is entirely subterranean.

Sources of water

The pavilions are joined by terraces that drain through the plinth walls and seem to have been originally free from soil and vegetation (Fig 3.15). Now they are mostly silted up (Fig 3.16) and termites are attacking the bases of the posts, the lower rails and the bottoms of the balusters.

Fig 3.15
The terraces between pavilions were originally free from soil and vegetation. Water from the roofs flowed down to stone-slat covered troughs.

Fig 3.16
Some of the terraces are now filled with soil and vegetation.

The old metal gutters around the main house are now distorted and leaking badly (Fig 3.17) and rainwater pipes are defective (Fig 3.18). These problems cause localised decay to modify the timber (*see* Chapter 4) and allow consequent termite damage.

Fig 3.17
Water drips down from a leaking gutter onto the top of the steps. The decking boards and the balustrade have been destroyed within the zone of wetting.

Fig 3.18
The bottom sections of rainwater pipes have been lost, allowing water to saturate the walls and termites to invade the lower section of the balustrade.

Pest management

Clearing the terraces and replacing the drainage system should be the first undertaking and may be sufficient to control the termites, although localised treatment may be necessary. Replacement timbers should be termite resistant or pretreated. As much as possible of the original timber should be conserved, and useable removed sections should be stored above ground in as dry an environment as is possible.

Case Study 4: Shwe-Kyaung Monastery, Mandalay, Myanmar

History

This monastery (Fig 3.19) was originally the royal apartment of King Mindon Min from the palace at Amarapura. The capital became Mandalay and the palace was moved in 1878. King Mindon Min died a few years later and his son thought that the building was haunted by the spirit

Fig 3.19
The monastery is constructed
from teak, on 150 teak
columns.

Fig 3.20
Rainwater drains from the
roofs down onto the balcony
where it drips between the
boards.

Fig 3.21
Rainwater drains down
through the balcony boards
onto a solid sloping apron,
which is now cracked. The
water should be diverted
around the outer columns
by low surrounds around the
column bases, but these are
now defective.

of his father and so it was removed outside the palace precinct and
reconstructed as a monastery. The remainder of the palace was destroyed
during World War II so that this is the only part remaining.

Construction and wood

The monastery floor area is reached by five sets of stone steps and is
constructed on 150 teak columns. The outer columns extend above the
balcony and are capped with stone. Rainwater from the roofs would
percolate through the balcony floor (Fig 3.20) and onto the concrete
apron below. From there it should drain away between the outer columns,
which were protected by low circular surrounds (Fig 3.21).

Termites

The termites we found attacking the outer columns were *Hypotermes xenotermitis*. This is a small genus of fungus-growing termites belonging to the Termitidae. They may produce low mounds, or nests may be entirely subterranean.

There were also occasional heaps of pellets at the bases of inner columns indicating the presence of drywood termites (Plate 10). These were probably a species of the genus *Cryptotermes*, but the termites could not be found without damaging the columns.

Sources of moisture

The caps of the outer columns had become displaced, allowing water into the column heads. This had allowed fungi to become established and termites had colonised wood that is both juvenile core wood and partially decayed (Fig 3.22 and *see* Chapter 4).

Fig 3.22
Fungus and termites have destroyed the top of the column under the displaced cap (a). The termites are able to forage down through the juvenile core (b).

Below the balcony, the concrete apron had cracked and the surroundings of the column bases had crumbled (*see* Fig 3.21). Water percolating down from above or blowing in from the sides would enter the cracks and drying would be impeded by the concrete. This trapped moisture had allowed fungi to colonise the column bases, followed by termites.

The stone steps were also cracked and damaged so that there was wet masonry against the timber structure (Fig 3.23).

Fig 3.23
Cracks at the top of the stone steps allow water to penetrate down into the subfloor area below.

Fig 3.24
Balcony timbers embedded in the wet stonework beneath the steps have been attacked by a poroid decay fungus. This allows termites to invade the teak joists.

This has resulted in the growth of a decay fungus (Fig 3.24) and consequent termite damage.

Pest management

Many of the outer columns required extensive repair or replacement. The primary consideration was then to make the drainage functional again so that there was no combination of decay and termites.

The important point to be made by this case study is that the termites had only damaged the timbers where sustained water penetration had caused them to decay (*see* Chapter 4). The remaining enormous volume of teak timbers in the building was not at risk from termite infestation and treatment was not required. Remedial works could be focused on the problem without unnecessary interventions.

References

Abdurahman, A 1990 'Foraging activity and control of termites in western Ethiopia'. Unpublished PhD Thesis, University of London

Al-Zadjali, T S, Abd-Allah, F S and El-Haidari, H S 2006 'Insect pests attacking date palms and dates in Sultanate of Oman'. *Egypt J Agric Res* **84** (1), 51–9

Badawi, A, Al-Kady, A L and Faragalla, A A 1986 'Termites (Isoptera) of Saudi Arabia, their hosts and geographical distribution'. *J App Entomol* **101** (1–5), 413–20

Chiu, C-I, Mullins, A J, Kuan, K-C, Lin, M-D, Su, N-Y and Li, H-F 2021 'Termite salinity tolerance and potential for transoceanic dispersal through rafting'. *Ecol Entomol* **46** (1), 106–16

Debelo, D G and Degaga, E G 2014 'Preliminary studies on termite damage on rural houses in the Central Rift Valley of Ethiopia'. *Afr J Agric Res* **9** (39), 2901–10

Gleeson, C F and Poulin, R 1989 'Gold exploration in Niger using soils and termitaria'. *J Geo-Chem Explor* **31** (3), 253–83

Harris, W V 1961 *Termites: Their Recognition and Control*. London: Longmans

Hocking, B 1965 'Notes on some African termites'. *Proc R Ent Soc Lond Series A* **40**, 83–7

Korb, J 2011 'Termite mound architecture, from function to construction' *in* Bignell, D E, Roisin, Y and Lo, N (eds) *Biology of Termites: A Modern Synthesis*. London: Springer, 349–73

Sands, W A 1992 'The termite genus *Amitermes* in Africa and the Middle East'. *NRI Bull* **51**

Steward, R C 1981 'The temperature preferences and climatic adaptations of building damaging dry-wood termites (*Cryptotermes*: Isoptera)'. *J Therm Biol* **6** (3), 153–60

Steward, R C 1982 'Comparison of the behavioural and physiological responses to humidity of five species of dry-wood termites *Cryptotermes* species'. *Physiol Entomol* **7**, 71–82

Turner, J S 2001 'On the mound of *Macrotermes michaelseni* as an organ of respiratory gas exchange'. *Physiol Biochem Zool* **74** (6), 798–822

Turner, J S 2006 'Termites as mediators of the water economy of arid savanna ecosystems' *in* D'Odorico, P, Porporato, A and Wilkinson Runyan, C (eds) *Dryland Ecohydrology*. Cham: Springer, 303–13.

West, W F 1970 'Termite prospecting'. *Chamber Mines J* **12** (10), 32–5

Wigglesworth, V B 1985 'The transfer of lipid in insects from the epidermal cells to the cuticle', *Tissue Cell* **17** (2), 249–65

Williams, R M C, Becker, G and Liese, W 1976 'Factors limiting the distribution of building-damaging dry-wood termites (Isoptera, *Cryptotermes* spp.) in Africa'. *Mater Org* **3** suppl, 393–406

Wiltz, B 2015 'Effect of temperature and humidity on survival of *Coptotermes formosanus* and *Reticulitermes flavipes* (Isoptera: Rhinotermitidae)'. *Sociobiol* **59** (2), 391–4

Woodrow, R J, Grace, J K, Nelson, L J and Haverty, M I 2000 'Modification of cuticular hydrocarbons of *Cryptotermes brevis* (Isoptera: Kalotermitidae) in response to temperature and relative humidity'. *Environ Entomol* **29** (6), 1100–7, https://doi.org/10.1603/0046-225X-29.6.1100

Yarwood, D J 1999 'Traditional building construction in an historic Arabian town'. *Constr Hist* **15**, 57–77

Zukowski, J and Su, N-Y 2020 'A comparison of morphology among four termite species with different moisture requirements'. *Insects* **11** (5), 262, https://doi.org/10.3390/insects11050262

4 Wood, durability and termites

4.1 Food resources

All termites have been assigned to four groups depending on their food resource (Donovan *et al* 2001). These are considered to demonstrate an increasing humification gradient from living plants to dispersed organic material in soil.

Group i: lower termites that feed on dead wood, grass.
Group ii: higher termites (Termitidae) feeding on a range of materials including dead wood, grass, leaf or small wood litter, and lichen, mosses or algae on bark. They differ from Group i in their gut symbionts.
Group iii: higher termites feeding in the upper layers of the soil that are rich in organic materials at the plant-soil interface.
Group iv: higher termites digesting particulate plant material in mineral soil. These are soil feeders.

Groups i and ii contain about 50% of termite species, and it is termites within these groups that are a concern for building conservation.

4.2 Wood and termite digestion

4.2.1 What is wood?

Wood (xylem) is made up of cells which are constructed from three carbon-based structural polymers. An organism's ability to use wood as a food resource is dependent on its ability to break these down together with any metabolised cell contents, or to utilise an intermediate organism that will accomplish this task.

The following polymers are present, roughly in the percentages given.

1 Cellulose (40%–50%)
This is formed from about 7,000–15,000 glucose sugar molecules bonded together.

Cellulose molecules form long bundles called microfibrils within the cell walls of plants. Those within the thickest layer of wall in mature wood are orientated so that they run approximately parallel to the cell axis and therefore up the tree. Large areas of the microfibrils are more or less crystalline and resist degradation. The remaining areas are termed amorphous.

2 Hemicelluloses (20%–30%)
Glucose forms a linear polymer, but hemicelluloses are polymers made from several different sugars, about 500 to 3,000 units long, that are cross

linked. They have a random structure with little strength and are easily broken down. They surround the cellulose fibres and cross link with them to provide extra support.

3 Lignin (20%–40%)

Lignin is a complex material formed as a secondary compound from different combinations of three phenolic (phenylpropane) units. It encrusts the cellulose/hemicelluloses, forming an insoluble 'lignocellulose' matrix. Lignin provides rigidity, and it is also hydrophobic so that it can line conducting cells, allowing the tree's vascular system to conduct efficiently. Degradation of lignin presents problems because of the variety of linkages between the units. It therefore acts as an important barrier to pests and pathogens. Lignin deconstruction becomes the first necessary step to expose the hemicelluloses and cellulose for digestion.

Plant biomass (lignocellulose) is the most plentiful renewable resource on earth, with the potential to replace petroleum-based raw materials and provide energy from biofuels. Termites are some of the most efficient decomposers of the lignocellulose complex, and their gut microorganism assemblages (microbiomes) are making them of considerable interest in potential 'bioreactors' (Auer *et al* 2017).

4.2.2 Termite digestion

Termites include species that feed on organic material ranging from living trees, dry grass, decayed wood, leaf litter and organic residues in soil. The digestion of cellulose and hemicelluloses is efficiently accomplished, but lignin presents a challenge and only modifications of the molecule seem to occur so that it is mostly excreted in termite faeces and frequently used as a construction material. It is also used to line galleries (Plate 13), where it may be further metabolised by microbes and then re-ingested by the termites.

In wood-feeding worker termites, ingested material is efficiently ground (triturated) by the molar surface of the mandibles and then passed via the oesophagus and crop to the gizzard (Fig 4.1). The gizzard is muscular and contains thick folds of cuticle (cuticular armature) that continue the breakdown of the food. Soluble compounds are absorbed by the mid gut, but digestion in the lower termites is aided by single-celled organisms (flagellate parabasalid and oxymonadid protozoa with bacteria) which are contained mostly in an enlarged area of the hind gut known as the paunch. It has generally been assumed that the termites could not survive without these protists, but it is now known that termites also produce the required digestive enzymes (cellulases) to break down at least the amorphous regions of the cellulose microfibrils, although the production of enzymes to break down their enshrouding hemicellulose polymers seems still uncertain. The relationship between the termites and their gut protists is therefore less clear (Lo *et al* 2011). The gut structures found in *Coptotermes* are shown in Fig 4.1, but there is some variation depending on diet.

Higher termites do not have protists in their gut systems, and their function is undertaken by a wide range of bacteria. The limitations on lignin breakdown remain the same, but the Macrotermitidae incorporate

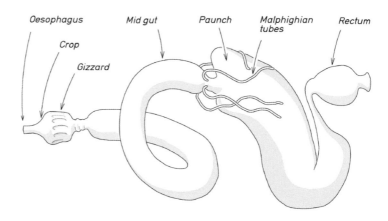

Oesophagus   Mid gut   Paunch   Malphighian tubes   Rectum

Crop

Gizzard

Coptotermes (Rhinotermitidae)

Fig 4.1
The digestive system of a
worker lower termite.
[© Iain McCaig]

faeces into fungus combs (Fig 4.2) and these fungi (*Termitomyces*: Basidiomycetes) are able to digest lignin. The termites then feed on the fungus asexual spores (conidia) and ageing mycelium. This is essentially a composting process regulated by the termites.

The fungus spores are rich in enzymes and the mycelium contains a significantly elevated nitrogen content. The latter is important because

Fig 4.2
Fungus comb of *Odontotermes*
***sp.* (Jima, Ethiopia).**

nitrogen is required to make amino acids/proteins and the amount of nitrogen present in dead woody tissue is very small (0.03–0.10%). However, wood/leaf litter-feeding termite families also have an ability to fix atmospheric nitrogen via spirochete bacteria in their hind gut. This ability contributes significant amounts of nitrogen to forest soils (Curtis and Waller 1998).

Extraneous sources of nitrogen may also make wood more favourable, and Hadlington and Beck (1996) observed that poles which dogs urinated against were more colonised by termites than those to which the dogs did not have access. A similar effect is found in the UK where old furniture joints and fixing blocks that were bonded with animal glue, or floorboards below bat roosts, may all become heavily infested with furniture beetle (*Anobium punctum*) because of their artificially elevated nitrogen content (Ridout 2000).

An effective integrated scheme for termite management requires an understanding of tree growth and both the origin and distribution of durability in timber.

4.3 Trees, growth and durability

4.3.1 Softwoods and hardwoods
Commercial timbers are grouped as either softwoods or hardwoods. The terms were originally used in England to differentiate between traditional hard building timbers and the imported soft, easily worked, wood more suitable for interior use. Technically, the hardwood was timber from broad-leafed trees (oak, ash etc) while the softwood was from cone-bearing trees (pine, spruce etc). The terminology remains in use but can be misleading, since, for example, balsa wood would be classed as a hardwood in spite of the fact that it is extremely soft. Botanically, the conifers (gymnosperms) are more primitive and first appear in the fossil record about 350 million years ago. The fossil record for the flowering plants, including the flowering trees (angiosperms), extends no further back than 100 million years when they are thought to have diverged rapidly from some unknown gymnosperm ancestor. This difference in evolution is reflected in the structure and chemistry of the wood and sometimes in the organisms that cause decay.

4.3.2 Conduction and strength
According to the cohesion–tension theory, water moves from the roots to the crown of a tree under negative pressure because of transpiration from the leaves (Fig 4.3). The woody stem has therefore to provide both strength to support the leafy crown and some form of conduction system. The theory also requires a continuous stream of water within the conduction system because any gas bubbles that formed would rapidly expand under the negative pressure and halt the hydraulic conductivity.

The majority of cells in softwood are called tracheids. These are hollow needle-shaped units that are packed closely together and in cross section resemble a honeycomb. Tracheids produced at the beginning of the growing season (early wood) are thin walled so that the interior cell volume is maximised for conduction. Those produced at the other end

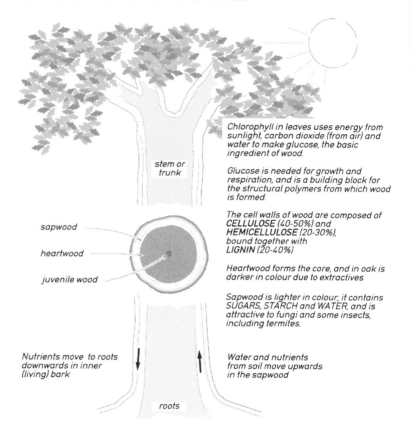

SIMPLIFIED ANATOMY OF AN OAK TREE

Chlorophyll in leaves uses energy from sunlight, carbon dioxide (from air) and water to make glucose, the basic ingredient of wood.

Glucose is needed for growth and respiration, and is a building block for the structural polymers from which wood is formed.

The cell walls of wood are composed of CELLULOSE (40-50%) and HEMICELLULOSE (20-30%), bound together with LIGNIN (20-40%)

Heartwood forms the core, and in oak is darker in colour due to extractives

Sapwood is lighter in colour; it contains SUGARS, STARCH and WATER, and is attractive to fungi and some insects, including termites.

stem or trunk

sapwood

heartwood

juvenile wood

Nutrients move to roots downwards in inner (living) bark

Water and nutrients from soil move upwards in the sapwood

roots

Fig 4.3
Simplified anatomy of an oak tree.
[© Iain McCaig]

of the season (late wood) have a smaller radial diameter, are flatter, and have thicker walls to resist both the compression loads imposed by the crown and bending moments resulting from wind forces. Thick-walled tissue is usually darker and denser than thin walled, and the two together produce concentric annual rings, which may be distinctive. Tracheids are not open ended and conduction from cell to cell is via valves known as bordered pits. These valves protect against embolisms by trapping air bubbles and also help to protect against vascular pathogens.

In softwoods, conduction and strength are thus provided by changing the shape and wall thickness of one type of cell, but in hardwoods, cell types have become differentiated. The conduction cells, called vessels, are different to the supporting cells, known as fibres. Vessels are formed from stacks of long cylindrical cells with end walls that are digested so that they are more or less open or form perforation plates that trap gas bubbles. Strength is provided by long narrow cells called fibres, which may be thick walled or thin walled according to species and sometimes season. Fibres also may have valves of different kinds and assist with conduction.

Some hardwoods, such as oak and teak, are ring porous, producing wide vessels during spring to maximise water conduction to the new leaves, and this produces an easily discernible early wood. Conduction is less critical during the latter part of the year; the vessels produced then are much smaller, and the extra space is taken up with fibres: the result is late wood (Fig 4.4). Subterranean termites will preferentially attack early wood, leaving a distinctive laminar form of damage (Fig 4.5).

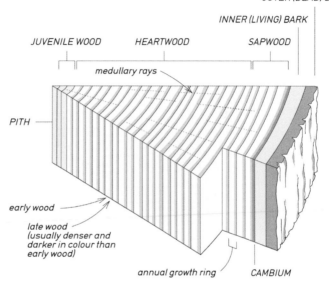

TYPICAL CROSS SECTION THROUGH TRUNK (NOT TO SCALE)

OUTER (DEAD) BARK

INNER (LIVING) BARK

JUVENILE WOOD   HEARTWOOD   SAPWOOD

medullary rays

PITH

early wood

late wood
(usually denser and
darker in colour than
early wood)

annual growth ring   CAMBIUM

Fig 4.4
Cross section through the trunk of a softwood tree such as pine or a ring-porous hardwood tree such as oak or teak.
[© Iain McCaig]

Fig 4.5
Typical damage caused by termites preferentially attacking early wood (Vietnam).

By contrast, diffuse-porous hardwood trees such as beech will produce similar amounts of vessels and fibres throughout the year so that the seasonal growth differences are not so easy to observe. Some hardwood species, such as walnut, are neither ring porous nor diffuse porous, but have intermediate cell constructions.

Growth of some tropical trees is continuous so that there are no seasonal markers, but growth can be interrupted by rainfall and other causes, though not on a regular basis.

The size and arrangement of hardwood pores together with other cell features enable species identification when mature wood is viewed as end grain.

4.3.3 Sapwood and heartwood

Growth progresses from a narrow layer of cell-producing tissue called the cambium, which is sandwiched between the bark and the underlying woody stem (*see* Fig 4.4). These cells originally contained a living material called protoplasm, but much of this is soon lost as the tree produces functional cells. Protoplasm then becomes restricted to blocks of tissue called parenchyma, which is particularly plentiful in the rays. Rays distribute water, minerals and other organic substances across the trunk.

The wood that is involved in active transportation and therefore contains living cells is called sapwood and this may comprise the entire thickness of a sapling. Sapwood also acts as a storage site for water and energy reserves such as starch. As the trunk increases in diameter, however, there becomes a time when the sapwood width is the maximum that the tree requires. Too little sapwood and the leaves will desiccate, too much and resources would be wasted – the system would not be efficient because living cells respire and consume energy. This maximum will depend on many factors, including tree species, growing location and interaction with other trees.

When optimum sapwood growth is reached, the cells at the centre of the trunk die, the nutrients they contained are relocated or converted into other substances, and heartwood production commences (*see* Plate 16). Heartwood no longer conducts water and therefore takes no further part in the transpiration stream of the tree. In some species the vessels become blocked with inclusions. As the years progress, the death of the inner sapwood cells continues to expand the heartwood so that it maintains the optimum sapwood content while assisting with structural support.

Sapwood, in the living tree, has a high moisture content with little air, making it a difficult environment for pathogens, and it also contains living material that can resist invading organisms by producing extra wood cells or surrounding them with toxic barriers. These barriers may be fungicides or gums and resins or other natural materials. The sapwood layer is therefore a protective outer layer. Most invading organisms must attack either via wounds in branches or roots that have caused the sapwood to dry, or via the central core, particularly after the tree becomes over-mature and begins to suffer changes associated with old age. It is in this way that ancient trees eventually become hollowed.

The heartwood may be dead, but it is not necessarily unprotected. Protection may come from non-structural compounds known as extractives (because they are mostly easily removed with solvents) that flood the inner sapwood cells as they die. It is extractives that give heartwood colour, odour and, in some tree species, durability.

Sapwood in the living tree provides protection, but this is lost when the tree is felled and the wood dries. If the wood has durability, then this is only in the heartwood provided that the wood chemistry is not modified by fungus and there are no changes associated with advanced age. Inclusions in conducting vessels and the blockage or incrustation of valves between conducting cells may also help protection in heartwood because they form a barrier to invading fungi.

This all means that the sapwood of any timber, be it pine or teak, will be inherently perishable because the extractives that provide durability have not been deposited. The innate durability of a building component constructed from these timbers is going to be significantly influenced by its sapwood content.

However, there may be some advantages in sapwood because its lack of extractives means that it can usually be easily impregnated with biocides, and high sapwood content is, therefore, not necessarily seen by the timber industry as a disadvantage. There are also times when durability is not the major consideration, and the contrast between sapwood and heartwood is exploited for decorative purposes.

The proportion of sapwood in a tree depends on many factors, but the spatial area of the heartwood and sapwood is dependent on growth-ring width. This, in turn, is influenced by the environment and silviculture (Plate 14). Gjerdrum (2003) notes that 'small crowns of pine provide a proportionally low potential for xylem production resulting in narrow rings and low sapwood production'. This suggests that trees grown within a forest where spread is restricted might have less sapwood that the trees of the same species growing around the periphery, but other factors such as the dominance of a tree within a stand will also have an effect.

Cedar (*Cedrus sp.*) has been used as a good termite-resistant timber and it has been shown (Güney 2018) that sapwood content was significantly correlated with trunk diameter (tree basal area). It was also correlated with crown area, thus supporting the work with pine.

The positive relationship between sapwood area and foliage biomass has been researched for two hardwood trees whose wood has a moderate to good resistance to termites. These are teak (*Tectona grandis*) and gamhar (*Gmelina arborea*). Both are grown as plantation trees and it is recommended that they be given plenty of space from an early age to rapidly increase diameter (Morataya *et al* 1999). This rapid increase requires greater water transport and therefore increases the volume of sapwood with no natural durability.

4.3.4 Juvenile wood, density and durability

The early years of tree growth produce juvenile wood, which will be sapwood that then converts into juvenile heartwood as the trunk expands. Juvenile wood differs from mature wood in several ways. The bundles of cellulose fibres in the cell walls are arranged at a shallower angle than in mature wood, the cells are shorter and there is more early wood so the wood is less dense. It is thought that these differences between juvenile and mature wood provide greater flexibility for the young branches in the leafy crown when juvenile, while stiffening the timber in the stem as the tree matures. The juvenile period may last for up to 20 years but seems commonly to be about 12 years. Juvenile wood may have poor strength and be more prone to movement (warping). The content of extractives that provide exploitable durability to the timber may also be lower in juvenile wood, and the extractive chemistry may differ from mature (Moya *et al* 2014). The juvenile core of a timber often makes an accessible route for termites (Fig 4.6) especially if there is an incipient heart rot.

Fig 4.6
This teak beam has a developing heart rot in its juvenile core and there are signs of termite activity in its centre.

Slow-growing timbers tend to be more dense, and denser timbers tend to be more durable, often because of their higher lignin content. Lignin is extremely resistant to some decay fungi, but others have developed the ability to degrade it, and any natural durability in timbers is provided by extractives. These, as discussed earlier, are produced when the living tissue (parenchyma) in the inner sapwood ray cells die and their stored starch and sugars are converted into fungitoxic compounds. Decay resistance in the trunks of large trees may increase from the centre to the sapwood and decrease up the stem.

Various authors have demonstrated that extractives removed from durable wood and added to perishable wood transfer the durability (Taylor *et al* 2002), but sometimes this is not a consequence of single compounds, and a range of moderately toxic compounds may act synergistically.

4.3.5 Forest growth and its consequences

The first trees to colonise a site are often classified as 'pioneer' species. These are shade-intolerant trees that grow rapidly but tend to have wood that is weaker and less decay resistant. They are also generally short lived because of branch breakage and stem decay. Growth of these pioneers may be a 'primary succession', in which case they have to overcome unfavourable conditions for establishment and their growth may be restricted, or a 'secondary succession' when the trees develop in land that has been cleared before, by nature or by human activity, and competition from existing vegetation has been reduced. These species may also grow within a forest if a tree falls and leaves a gap in the canopy. In this case the pioneer species will grow very rapidly towards the light.

Shade-intolerant trees may be slowly replaced with shade-tolerant trees, referred to as climax species or late-succession species, and these grow slowly in competition with their neighbours. It should be noted that shade intolerance and shade tolerance are not necessarily clear distinctions and many species fall between the two extremes.

Plantation silviculture provides the possibility to increase the yield of valuable timbers (although this may be at the cost of other properties), but climax species like the African mahoganies (*Khaya* spp. and *Entandophragma* spp.) are too slow growing and environmentally sensitive to exploit.

Some of the pioneer and intermediate trees are sufficiently tolerant of a range of environments and soil types that they make useful candidates for plantation growth. Teak (*Tectona grandis*) requires plenty of light to grow and has always been a valuable tree. It used to be harvested in India when the trees were about 120–150 years old, with a thinning cycle every 30 years to remove inferior or damaged trees and climbers (Pandey and Brown 2001). Current forestry practice, as recommended by these authors, is an initial extensive thinning at 4 to 5 years, a first thinning for timber production at 10 to 15 years, and a second at 15 to 20 years. Moya *et al* (2014) state that harvesting after 15 to 30 years when the logs have a diameter of between 12cm and 30cm is common. However, they note that carefully managed plantations on selected sites in other countries allow logs with a marketable diameter of 12cm to be produced in less than six years.

Heartwood production begins when the diameter is between 6cm and 10cm (Moya *et al* 2014) and the point to be made from these statistics is that marketed plantation timber on these short rotations is going to produce a timber with a high sapwood and juvenile wood content (Figs 4.7 and 4.8) with limited durability. Twelve years of juvenile growth that is predominantly sapwood is not particularly significant in a log from a 120-year-old tree, but it might make a considerable difference to the durability of timber that was harvested after 9 to 20 years. Figure 4.9 shows the distribution of sapwood when the log is converted.

A similar problem occurs with softwood. The most frequently used construction timber in the UK during the 19th century was Scots pine (*Pinus sylvestris*) imported from northern Europe, where a short growing season meant that trees reached a useful size for construction timbers in about 200 years. The growth period for the perishable sapwood was around 20 years and so the amount of sapwood present was insignificant. Natural durability was maximised and damage by wood-boring insects was minimal (Plates 14 and 15).

Unfortunately, wild-grown timber on a 200-year rotation was unsustainable. Secondary and plantation growth could produce a crop, where soils and climate were more favourable, in about 50 years (Ridout 2019). Growth rate could be enhanced by thinning to allow more space

Fig 4.7
Old forest-grown teak logs in a wood merchant's shop in Patiala (India). The narrow pale outer layer is sapwood.

Fig 4.8
Young plantation teak logs showing the pale remnants of wide sapwood bands. This was in the same merchant's shop as the logs in Figure 4.7. The distribution of sapwood is explained by Figure 4.9.

and thinnings could be sold. Thinning might occur at around 18 years and again at about 35 years (Plate 16). Now the 20 years of sapwood growth becomes significant, and durability can only be maximised by careful construction, good maintenance and, if these cannot be ensured, then impregnation with chemicals.

Large quantities of plantation-grown pine wood are exported and used for repairs in countries where termites may be a significant problem.

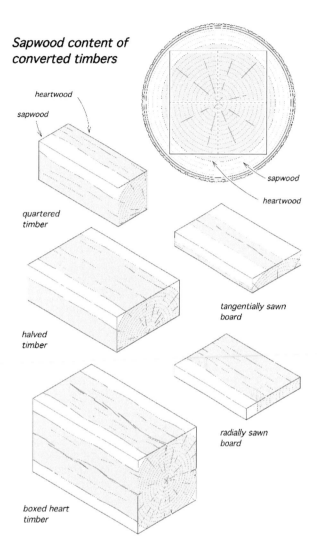

Sapwood content of converted timbers

heartwood

sapwood

sapwood

heartwood

quartered timber

halved timber

tangentially sawn board

radially sawn board

boxed heart timber

Fig 4.9
The distribution of sapwood in sawn components when a log is converted. The percentage of perishable sapwood in any particular species of timber will depend on the location where the tree was grown and the age when felled.
[© Iain McCaig]

4.4 Natural durability and fungi

Most possible food resources have a range of organisms with the ability to overcome adverse chemistry and, ultimately, adding biocides to timber may only alter the type of fungus that can invade. Coal-tar creosote, for example, is an excellent preservative because of the range of toxic compounds it contains, but impregnated timber may still be destroyed by the fungus *Neolentinus lepideus*. Wood-decay organisms tend to form a progression after initial barriers have been broken down, until nothing solid remains.

Durable timbers, therefore, only remain durable while their chemistry is intact. Red cedar (*Thuja plicata*) has a good natural durability because of an extractive called β-thujaplicin. The extractives in red cedar have been shown to provide resistance to termites (Taylor *et al* 2006), but there are fungi that will break them down if the wood remains wet enough for fungi to colonise. Lim *et al* (2005) investigated the buried ends of red cedar fence timbers and found that they were degraded in ground contact by 'pioneer' fungi that had a high tolerance for the extractive. Once the heartwood was detoxified, then other decay fungi could follow.

In 1995 English Heritage (now Historic England), together with Dutch and Irish partners, commenced a three-year investigation of the deathwatch beetle (*Xestobium rufovillosum*). This is an anobiid beetle that causes considerable damage to old oak structures in the UK and Europe. One of the tasks was to understand why the beetles could attack heartwood of normally durable oak after it had been colonised by fungi but could not invade when there was no fungus present. This research task was undertaken by TNO Building and Construction Research in Delft using highly sensitive analytical techniques (pyrolysis-direct chemical ionisation mass spectrometry followed by principal component-discriminant analysis). The findings were that the oak-rotting fungi tested caused radical changes to wood chemistry, even when the weight loss from decomposition was <3% and there were no visible signs of decay (Esser and Tass 2001).

It had previously been suggested that the role of the decay fungi was to soften the timber, which seemed unlikely to be the complete explanation because emerging deathwatch beetles are able to penetrate lead sheeting on roofs. A second suggestion was the fungi increased the nitrogen content of the wood. This was investigated by the protein/amino acid assay of infested and sound oak of different ages but could not be substantiated (Ridout and Ridout 2001).

Our conclusion was that sustained water penetration allows some form of fungus to develop. This, even from an early stage, will modify the chemistry of the wood, negating its natural durability and allowing wood-infesting insects to invade. The explanation is not unexpected, because many insects, including termites, selectively attack the dead parts of trees that have durable heartwood. A dead branch or trunk core will usually have been invaded by a fungus as part of the normal forest progression.

If a wood is termite resistant, then the insects may only consume the sapwood, which in a pole construction (Plates 17 and 18) means grazing across the sapwood surface. If persistent damp, such as a bearing in a damp wall, allows a fungus to invade, then the wood chemistry is modified and the bearing becomes a food resource for the termites (Plate 19).

If the chemistry of the wood has been modified, then it will remain vulnerable even after the moisture source that allowed the fungus to develop has been removed. Good maintenance then becomes essential.

Case Study 5: The Royal Tombs of Hue, Vietnam

History

In 1993 the Imperial City of Hue was declared a World Heritage Site. History has not been kind to Vietnam, and the temples and mausoleums were in a poor state of repair. The primary problem seemed to be a variety of subterranean termites. In 1995 UNESCO signed a partnership agreement with a major pesticide manufacturer in order to preserve and promote the world's historic treasures, and the termites at Hue were considered a suitable case for treatment. The task was not a simple one. They proposed the following schedule of works:

- Dismantle the mausoleum
- Treat the soil and walls
- Treat all new wood
- Restoration of the building and final treatment of all wood

This philosophy is well established for many types of decay organisms and in many countries. It is based on the idea that if an organism eats wood, and the building is made from wood, then the organism will eat the building.

The chemical formulations used were efficient insecticides based around an insect nerve poison, but co-ordination of the treatments with the restoration programme was difficult, the lingering smell of the insecticides was unpleasant, and the treatment programme was abandoned after the first temple had been dismantled and re-erected. A different opinion on the control of termites was requested.

Construction and wood

The starting point was the observation that, although the buildings were constructed from wood, many were 100–200 years old. If the termites could destroy these timbers, then the original buildings would have been lost many years ago. Clearly the two major timber species from which the buildings were constructed, *Erythrophloeum fordii* and *Hopea pierrei*, must be immune from termite attack unless something happens to change the properties of the wood. The comprehensive treatment of all timbers with chemicals will therefore serve no useful purpose.

Erythrophloeum fordii, commonly called iron wood (lim xanh), is an open-forest tree from South China and North Vietnam that has been extensively over-exploited because the wood is dense, durable and resistant to insect attack. It is capable of reaching a height of 37m–45m with a diameter of 200cm–250cm. This tree will grow in plantations, and a study of 45 plantations in Vietnam (Sein and Mitlöhner 2011) recorded an average height of about 12m (diameter 9cm) after 12 years and 15m (diameter 16cm) after 24 years. Trees are normally fertilised, pruned to provide straight stems, and thinned. Rotation is generally 20–25 years. These results suggest that a wild-grown forest tree, as used at Hue, with a diameter of 200cm (Fig 4.10) was several hundred years old when felled. The consequences of plantation growth for durability do not appear to have been investigated, but much of the timber must be juvenile wood or sapwood.

Fig 4.10
Searching for termites at the
Tomb of Dong Khanh, Hue,
Vietnam. The termites are
attacking the tops of the
lacquered columns where
roof leaks have allowed fungus
to become established in
termite-resistant timber.

Hopea pierrei known as merawan (kiền kiền) is a forest tree that
is now classified (IUCN Red List of Threatened Species 2011) as
endangered over most of Southeast Asia. It can reach a height of 40m
with a diameter of 180cm, but most are around 25m tall with a diameter
of about 50cm.

Termites
The Hue Monument Conservation Centre sent each type of termite
they found to Hanoi for investigation. The species collected are given in
Table 4.1.

Table 4.1 Termites identified for the Hue Monuments Conservation Centre

Family	Type	Identity
Kalotermitidae	Drywood	*Cryptotermes domesticus*
Rhinotermitidae	Subterranean	*Coptotermes formosanus**
	"	*Coptotermes travians**
	"	*Coptotermes curvignathus*
Termitidae	"	*Microtermes dimorphilus*
	"	*Odontotermes hainanensis*
	"	*Odontotermes formosanus*
	"	*Macrotermes aunandalei*
	"	*Nasutitermes parvonasutus*
	"	*Globitermes sulphureus**

*Species discussed by Harris (1971) as causing significant damage in buildings.

Sources of moisture

The two construction timbers have a good resistance to termites, but ongoing termite damage had developed where wood decay fungi (commonly *Phellinus senex*) had modified the wood chemistry (Fig 4.11). Decay was occurring because roof and drainage faults were allowing the timbers to remain wet for prolonged periods.

The relationship between fungi and termites had not previously been recognised and termites had therefore been seen as the primary decay

Fig 4.11
Fungus followed by termite damage promoted by a roof leak. The grey excrescence is a fruit of the poroid fungus *Phellinus senex*.

organisms rather than the secondary. The immediate consequences of this were that all timbers had been perceived to be vulnerable to termite attack and that extensive timber treatment and replacement had been considered essential.

Pest management

The treatment recommended by the pesticide manufacturer, while typical for normal situations, did not consider the particular problems at the mausoleum. The works schedule included the treatment of a large amount of interior timber that the termites could not attack unless it started to decay.

Fig 4.12
The consequences of reconstruction with a timber that has poor durability. The centre of this column has been destroyed by termites.

If termites are relegated to a secondary decay organism, then they become easier to control, because fungi, the primary colonisers, are heavily dependent on water. Good maintenance will prevent damage, and partially infected timbers may be repaired rather than replaced. The extensive use of biocides also becomes unnecessary, thus reducing the environmental impact of restoration works.

A second significant advantage of repair rather than replacement is that the quantity of new timber required is substantially reduced. Both *E. fordii* and *H. pierrei* have become expensive timbers and are increasingly difficult to obtain in Vietnam. We were informed that the conservation centres were only allowed to purchase timber from foreign companies with logging concessions; otherwise, the wood had to be imported illegally from Laos. Repair with alternative timbers has to be considered carefully. The temple of Thaito Mie in Hue, for example, was reconstructed in 1971 using timber that ultimately proved to have a poor durability (Fig 4.12). This was done because the traditional construction timbers were unavailable, but the result has been the destruction of the principal pillars by *Coptotermes formosanus*, one of the world's most destructive termites. The change in timber durability has changed the status of the termites from secondary to primary pest.

An accurate diagnosis of the problem allows the limited resources available to be targeted effectively. Manpower to undertake effective maintenance should be easier to obtain than massive amounts of insecticides and the environmental impact is diminished.

References

Auer, L, Lazuka, A, Sillam-Dussès, D, Miambi, E, O'Donohue, M and Hernandez-Raquet, G 2017 'Uncovering the potential of termite gut microbiome for lignocellulose bioconversion in anaerobic batch bioreactors'. *Front Microbiol* **8** article 2623, https://doi.org/10.3389/fmicb.2017.02623

Curtis, A D and Waller, D A 1998 'Seasonal patterns in nitrogen fixation in termites'. *Funct Ecol* **12** (5), 803–7

Donovan, S E, Eggleton, P and Bignell, D E 2001 'Gut content analysis and new feeding group classification of termites (Isoptera)'. *Ecol Entomol* **26** (4), 356–66

Esser, P and Tass, A 2001 'The roles of location, age and fungal decay in the chemical composition of oak' *in* Ridout, B (ed) *Timber: English Heritage Research Transactions* **4**, 79–86

Gjerdrum, P 2003 'Heartwood in relation to age and growth rate in *Pinus sylvestris* L. in Scandinavia'. *Forestry* **76**, 413–24

Güney, A 2018 'Sapwood area related to tree size, tree age and leaf area index in *Cedrus libani*'. *bilgesci* **2** (1), 83–91

Hadlington, P W and Beck, L 1996 *Australian Termites and Other Common Timber Pests*. Kensington, NSW: UNSW Press

Lim, Y W, Kim, J-J, Chedgy, R, Morris, P I and Breuil, C 2005 'Fungal diversity from western red cedar fences and their resistance to β-thujaplicin'. *Antonie Van Leeuwenhoek* **87**, 109–17, https://doi.org/10.1007/s10482-004-1729-x

Lo, N, Tokuda, G and Watanabe, H 2011 'Evolution and function of endogenous termite cellulases' *in* Bignell, D, Roisin, Y and Lo, N (eds) *Biology of Termites: A Modern Synthesis*, 51–67. Dordrecht: Springer

Morataya, R, Galloway, G, Berninger, F and Kanninan, M 1999 'Foliage biomass-sapwood (area and volume) relationships of *Tectona grandis* L.F. and *Gmelina arborea* Roxb.: silvicultural implications'. *For Ecol Manage* **113**, 231–9

Moya, A, Bond, B and Quesada, H 2014 'A review of heartwood properties of *Tectona grandis* trees from fast-growth plantations'. *Wood Sci Technol* **48**, 411–33

Pandey, D and Brown, C 2001 'Teak: a global overview'. *Unasylva* **51** (201), 3–13

Ridout, B 2000 *Timber Decay in Buildings: The Conservation Approach to Treatment*. London: English Heritage, Spon

Ridout, B 2019 *Timber Decay in Buildings and its Treatment*. Swindon: Historic England

Ridout, B and Ridout, E 2001 'The effect of fungi on the growth of deathwatch beetle larvae and their ability to attack oak' *in* Ridout, B (ed) 2001 *Timber: English Heritage Research Transactions* **4**, 87–91

Sein, C C and Mitlöhner, R 2011 *Erythrophloeum fordii Oliver: Ecology and silviculture in Vietnam*. Bogor, Indonesia: CIFOR

Taylor, A, Gartner, B and Morrell, J 2002 'Heartwood formation and natural durability – a review'. *Wood Fiber Sci* **34** (4), 587–611

Taylor, A M, Gartner, B L and Morrell, J J 2006 'Effects of heartwood extractive fractions of *Thuja plicata* and *Chamaecyparis nootkatensison* wood degradation by termites or fungi'. *J Wood Sci* **52**, 147–53

5 A review of the most destructive termites

5.1 The problem

As noted in Chapter 1, there are approximately 2,600 species of termite worldwide. But only a few hundred species of termite impinge on human activities sufficiently to be considered important pests, and of these only a few pose a significant threat to structures and buildings.

But while the number of destructive species may be small, the number of individual insects belonging to them is vast. Grace (1992), reporting on mark-release-recapture investigations using fat-soluble dyes in the USA, found that colonies of *Coptotermes formosanus*, *Reticulitermes flavipes* and *Heterotermes aureus* could contain from several hundred thousand to several million termites. A colony might forage over an area exceeding 3,000m^2 and consume up to 1kg of wood each day. Populations of this size become exceedingly difficult to eradicate.

This chapter focuses on the termite genera that are potentially most harmful to heritage buildings and provides information that will help the reader to identify them, wherever they might be found (identification of individual species requires specialist investigation and is beyond the scope of this book). Identification is important because the policy for managing an infestation will be influenced by genera-specific parameters including moisture requirement (drywood, dampwood, subterranean), potential to produce satellite nests in buildings, and potential colony size.

Harris (1961) produced a world list of termites that damage buildings and recorded 109 species, of which 53 were classed as major pests. Edwards and Mill (1986) added a few more and then the list was updated from the literature by Krishna *et al* (2013). They listed 231 species that had been reported from building timbers and gave 87 serious pest status. There are a few anomalies, so the genera included in this chapter are based on all three lists. A few less destructive species have also been added where their extensive geographical distribution or changing habitat makes them of potential interest.

5.2 Distribution

Termites are found on every continent except Antarctica, but the majority are restricted to tropical, subtropical and warm temperate regions. The greatest diversity is found in the rainforests of Central Africa. Human activities and climate change, however, may alter distribution patterns and the following categories have been defined (Evans *et al* 2013):

- **Native or natural:** the original community of species of an area with no or little human modification.
- **Introduced:** a species outside its native range due to human transport.

- **Invasive:** the expansion of an established non-native
  population outside its native range.

A native species that has not caused much damage may become a serious pest in buildings if it is provided with a more favourable environment, which might include a moisture source and a perishable native or introduced timber. For example, *Reticulitermes balkanensis* is fairly common under discarded remnants of timber on the Greek island of Skopelos, but there are few signs that they are invading buildings at present. However, this may change with the introduction of new forms of construction.

Introduced termite species tend to remain restricted to the type of modified environment into which they were introduced. This may be structures or cultivated crops where they may cause considerable damage. The *Reticulitermes* infestation in Devon, UK, is an example of a localised accidental introduction.

By 2013, 28 species had been classed as invasive and therefore the greatest potential threat to heritage buildings. These are all lower termite wood feeders and wood nesters, and share the ability to form new small colonies that might be easily transportable. For example, a paperback book taken from a shelf in a home in California was found to have a single entry hole on the top. Inside was a short gallery containing a pair of drywood termites (*Incisitermes minor*) that had shed their wings and were preparing to found a colony (Gulmahamad 1997). A study of gallery formation in hardwood shipping pallets found up to eight colonies of the drywood species *Cryptotermes brevis* in individual items (Grace *et al* 2009).

These reproductively viable 'propagules' form most easily in termite genera where all individuals, with the exception of soldiers, are capable of becoming secondary reproductives (ergatoid reproductives).

The current distribution of invasive species will certainly increase in response to climate change, accelerating urbanisation and global trade (Buczkowski and Bertelsmeier 2017). Invasive organisms are, by nature, able to accommodate a range of environments and to outcompete less vigorous species. There is also the risk of hybrids that may be more invasive. It has been reported that two of the world's most destructive and economically important termites, *Coptotermes formosanus* and *Coptotermes gestroi*, are hybridising in Florida to produce colonies with twice the normal growth rate (Chouvenc *et al* 2015).

The following keys to alates and soldiers should allow most termites that are found to be allocated to a family. The generic information that follows the keys contains information on the termites that cause most of the damage to buildings around the world and it is hoped that a brief description and drawing of a typical soldier's head, the part with the most useful distinguishing characteristics, will aid identification.

A few species are noted under each genus to illustrate geographical range and habitats, but these lists are only intended as a guide to further investigations if necessary and cannot be inclusive. There are many termite genera and species that may less commonly invade buildings if their natural environmental requirements are replicated.

5.3 Keys to termite families

'Dichotomous keys' are used extensively by biologists for the identification of organisms at different levels of classification. The following keys are for the identification of termite families based on alates and soldiers. The reader starts at couplet 1 and has to decide at each stage (for example 1a or 1b) which of the two descriptions best fits the specimen being examined. At the end of each description is the number of the next couplet to be considered or the family name for the termite. The details discussed in each couplet are illustrated by a drawing and the external anatomy of a typical soldier and king or queen (wings shed) is shown in Fig 5.1 to locate the structures discussed. The key is reproduced (slightly updated) with the permission of Rentokil Initial plc from *Termites in Buildings: Their Biology and Control* (Rentokil Library) Edwards and Mill, 1986 © Rentokil Initial plc.

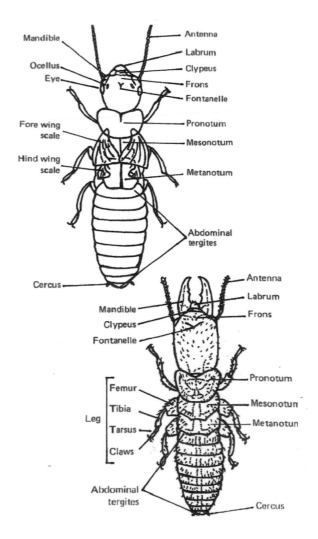

Fig 5.1
The external anatomy of a king or queen termite (wings shed) and a soldier.

Key I: Alates (kings and queens)

1a   Hind wing with an anal lobe (a);   Mastotermitidae
tarsi with five distinct segments (b);
fontanelle absent (c); ocelli present (d).
Forewing with full venation, the radius
branched (e). Left mandible with first
and third marginal teeth prominent,
but with the second flattened to form a
cutting edge (f). In tropical Australia,
introduced into Papua New Guinea.

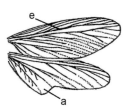 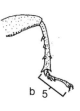 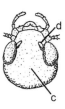 

1b   Hind wing without anal lobe   2
(a); tarsi four segmented (b) or
apparently five with the first more
or less divided (c).

 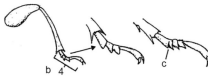

2a   Tarsi four segmented.   3
2b   Tarsi apparently five segmented   Archotermopsidae
        with the first more or less divided.

3a   Fontanelle absent (a); postclypeus   4
flat, without central groove or
distinct hind margin (b); not
clearly separated from anteclypeus
(c). Forewing with distinct costa
(S), radius (R) and radial sector
(Rs), which is usually branched.

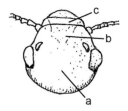 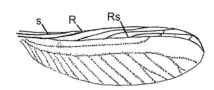

3b   Fontanelle present (a) though   5
     sometimes difficult to see.
     Postclypeus margined with a
     central groove (b); separated from
     anteclypeus (c) and often raised
     above frons. Forewing with front
     veins fused into a thick radial sector
     (Rs) behind the front edge.

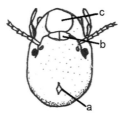 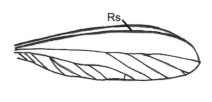

4a   Ocelli present (a); cerci with two   Kalotermitidae
     segments (b); left mandible with
     two marginal teeth (c); forewing
     scales (d) much larger than hind
     (e).

4b   Ocelli absent (a); cerci with three   Hodotermitidae
     to eight segments (b); left mandible
     with two or three marginal teeth
     (c); forewing scales (d) a similar
     length to hind (e).

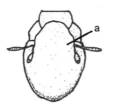  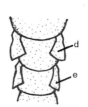

5a   Left mandible with three prominent   Rhinotermitidae
marginal teeth (a); forewing scales
(b) much larger than hind (c)
(except in the Psammotermitinae of
arid Africa and the Middle East).

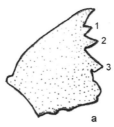

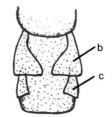

5b   Left mandible with one (a) or two   6
(b) prominent marginal teeth.

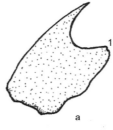

6a   Left mandible with only one   Serritermitidae
prominent marginal tooth which is
widely separated from the relatively
huge apical tooth: the second and
third marginals virtually obsolete
(a); forewing scales (b) much larger
than hind (c). In central South
America only.

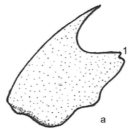

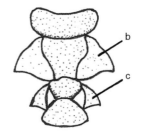

| 6b | Left mandible with first and third marginal teeth prominent, the second flattened to a cutting edge and not always distinguishable (a); forewing scale (b) about the same size as hind (c) and widely separated from its base. | Termitidae |

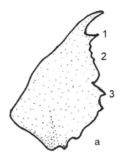
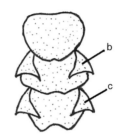

Key 2: Soldiers

| 1a | Tarsi five segmented (a); fontanelle absent (b); pronotum as wide as head and saddle shaped (c); cerci five-segmented (d); antennae with 20 or more segments. In tropical Australia, introduced into Papua New Guinea. | Mastotermitidae |

| 1b | Tarsi four segmented (a), or apparently five with the first more or less divided (b). | 2 |

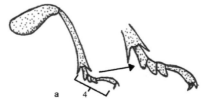

| 2a | Tarsi four segmented or if the first partly divided then eyes prominent and pigmented. | 3 |
| 2b | Tarsi apparently five segmented with the first more or less divided. No pigmented eyes. Pronotum small, narrower than head. | Archotermopsidae |

| 3a | Fontanelle absent (a); pronotum with front margin concave (b); pronotum flat (c) or if saddle shaped (d), then head with well-pigmented eyes (e). | 4 |

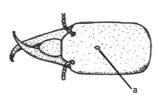

| 3b | Fontanelle present (a) but undetectable in some Termitidae. In these, the pronotum is saddle shaped (b) and the head without eyes (c). | 5 |

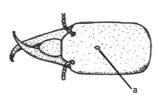
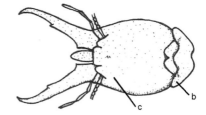

| 4a | Cerci with two segments (a); antennae with 19 or fewer segments; tarsi always distinctly four segmented (b); eyes absent or rudimentary and generally unpigmented. | Kalotermitidae |

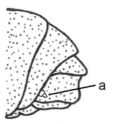

4b   Cerci with three to eight segments   Hodotermitidae
      (a); antennae usually with more
      than 22 segments; tarsi usually four
      segmented, but first sometimes
      partly divided (b); eyes prominent
      and pigmented.

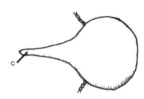

5a   Pronotum saddle-shaped with   Termitidae
      anterior lobes (a); head without
      eyes (b); head in a few species
      round with fontanelle at the end of a
      tubular projection (c).

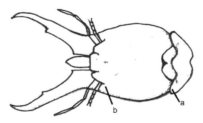

5b   Pronotum flat, without anterior   6
      lobes (a); front margin straight (b),
      concave or drawn out to a point only
      in the middle.

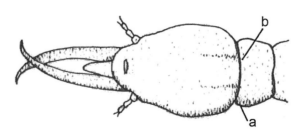

6a	Mandibles straight and coarsely serrated along most of their inner margins (a). In central South America only.	Serritermitidae

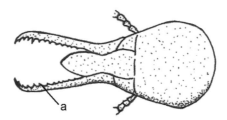

6b	Mandibles without teeth (a) or, if toothed, never coarsely serrated over most of their length: tips usually well curved if with many teeth (b).	Rhinotermitidae

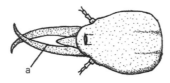 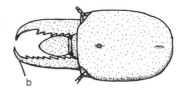

5.4 Families that include the most destructive genera

This section gives details of the termite genera that are most damaging to structures and buildings. The shape and structure of soldier termite heads are described as an aid to identification. Alates also have important distinguishing features, but swarming is seasonal, and the termites observed might not be the same as the ones invading building timbers. Therefore, soldier termites provide a more assured means of identification, even though they may be uncommon in some genera.

5.4.1 Family: Mastotermitidae
The head is round with short and toothed mandibles. The antennae have 20–26 segments. The length is normally 11mm–13mm.

Small subterranean nests are formed in hollowed-out old tree stumps, posts or poles and colony spread is mostly by neotenics (kings and queens developing from nymphs or worker caste, not from alates). Natural colonies are generally quite small, but a network of nests connected by subterranean tunnels may form in areas where there is extensive horticulture or forestry.

Mastotermes
There is only one living species in the genus.

Mastotermes darwiniensis: Native to tropical Australia and now invasive in Papua New Guinea where it infests both buildings and trees (Fig 5.2).

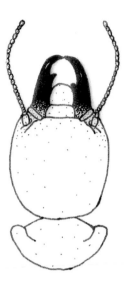

Fig 5.2
Head and pronotum of
Mastotermes darwiniensis
soldier.

5.4.2 Family: Kalotermitidae
Soldiers may have rudimentary eyes which are generally unpigmented. The left mandible has a tooth at the apex (apical tooth) and three on the inner edge (marginal teeth), but the latter may be considerably reduced in prominence. The right mandible has an apical tooth and two marginal teeth, but again the marginals may be so reduced as to be nearly absent. The pronotum is as wide or wider than the head. The cerci are short with two segments.

These are small-colony termites that colonise single pieces of timber without the need for ground contact, and are easily transported accidentally around the world. There are about 500 extant species in 21 genera that mostly live in the dead parts of trees. These are generalised wood feeders and their distribution is governed by habitat and climate rather than by host wood preference. Three genera contain species that are listed as significant building pests. *Kalotermes* is also included here because damage, though usually restricted, is common in heritage buildings over a wide area of the Mediterranean region.

Cryptotermes
Cryptotermes is the third-largest genus in the family with around 70 species distributed throughout the tropics. Soldiers are epiphragmotic (Fig 5.3). This means that they have plug-shaped heads, nearly as wide as long, that can be used to seal galleries against predators (Fig 5.4). Mandibles are short compared to the length of the head capsule. Soldiers are mostly 4mm–5mm long and antennae have 16–19 segments.

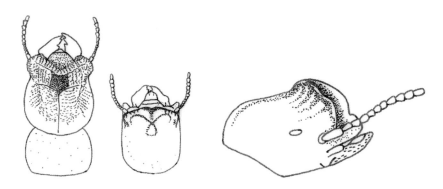

Fig 5.3
Heads of *Cryptotermes spp.*
soldiers, *C. brevis* including the
pronotum, and *C. domesticus.*
[Redrawn from **CSIRO
Bulletin 286** © **CSIRO
Australia**]

Fig 5.4
Epiphragmotic head of
Cryptotermes cynocephalus as
seen from the side.
[Redrawn from **CSIRO
Bulletin 286** © **CSIRO
Australia**]

Colonies rarely exceed a few hundred individuals, but nymphs readily develop into neotenics so that fragmented colonies are easily transported and become viable. Primary reproductives are not replaced unless they die, and colonies tend to die out after around 10 years as egg production by the ageing queen declines.

Many species are indigenous in different countries, but a few have been spread around the world by commerce and other human activities. This spread has been assisted by the insect's apparent preference for coastal environments, and a few species have been distributed so widely that they have become serious pests of buildings. For example, a study of the genus in the West Indies (Scheffrahn and Krecek 1999) found 17 indigenous species and 4 non-indigenous that had been spread by maritime activities probably dating back to the early slave trade.

The following list includes those species and others of particular significance.

Cryptotermes bengalensis: Bangladesh, India, Thailand and introduced into Sri Lanka. It attacks woodwork and trees.

Cryptotermes brevis (West Indian drywood termite): Now distributed across most of the tropical and subtropical regions around the world. It seems to have originated in the coastal deserts of Peru and Chile and is mostly found associated with man-made structures and wooden items. Scheffrahn (2019) considers it to be the most destructive and widespread pest in the New World (*see* Figs 5.3 and 5.4).

Cryptotermes cynocephalus: Malaysia, Philippines; now found in Australia and Hawaii. It is uncommon in its natural habitat and is mostly found nesting in buildings.

Cryptotermes declivis: Southern China, where it is a pest of structural timbers.

Cryptotermes domesticus: Native to Southeast Asia, China, Pacific Islands and Australia, where it is considered to be a minor pest of structural timber and furniture (*see* Fig 5.3).

Cryptotermes dudleyi: Another species that probably originated in the Indo-Malayan region, but has now been transported to the West Indies, South America and Africa. It is mostly found infesting houses.

Cryptotermes havilandi: Native to Africa and now widespread in the West Indies where it seems to be more commonly associated with the dead parts of standing trees than buildings. It is an important structural pest in West Africa.

Incisitermes

There are about 30 living species in this genus, distributed from the Far East and Australia across the Pacific to the eastern side of South America and the southern states of the USA. Soldiers' head capsules are more or less rectangular and mandibles are the stout and toothed crushing kind. The front margin of the pronotum is deeply concave or incised and the third antennal segment is frequently enlarged. Soldiers are about 6mm–10mm long with 15–20 antennal segments.

Colonies are restricted to single pieces of wood or pieces pressed closely together, and they are generally rather larger than those of *Cryptotermes*, reaching perhaps 2,000 individuals. A colony grows slowly, and a detailed study of *I. minor* (Harvey 1934) found that after two years a colony still only consisted of a king, a queen, a soldier and about a dozen nymphs. However, a well-established colony then develops rapidly and may contain over 1,000 individuals after about seven years. Such a colony, under optimum conditions (26°C–32°C) would consume about 0.25kg of wood each day.

Their natural environment is dead parts of trees, both standing and fallen.

Incisitermes marginipennis: Guatemala and Mexico, where it attacks building timbers.

Incisitermes minor (western drywood termite): Mainly in California, where it is a particular pest in wood-frame construction. It is easily transported in infested goods and infested boats. There are reports from many other states and even from a heavily infested structure in Toronto, Canada. Soldiers are 8mm–10mm long (Fig 5.5).

Fig 5.5

Head and pronotum of an
***Incisitermes minor* soldier.**

Incisitermes schwarzi: Common in southern coastal Florida, where it attacks utility poles and woodwork in buildings.

Incisitermes snyderi (light south-eastern drywood termite): The soldiers are about 7mm–10mm long and their heads are more yellow than the dark brown *I. minor*. This is the only endemic drywood termite in the south-east of the USA that commonly infests structures. However, colonies do not seem to be so easily transported and established in new locations as *I. minor*.

Kalotermes

This genus contains around 20 species found within temperate and subtropical zones. The shape of the soldiers is generally similar to those of *Incisitermes*, but the anterior margin of the pronotum is more shallowly concave. The mandibles sometimes have a basal hump on the outer margin. Soldiers are about 7mm–12mm long, depending on species.

K. flavicollis larvae and nymphs have been shown to readily differentiate into other castes if colony regeneration becomes necessary and become neotenic reproductives without changing their juvenile form (Harris 1970). The antennae have 16–19 segments.

Kalotermes approximatus (dark southern drywood termite): The only *Kalotermes* species found in the USA and it is distributed around Florida and other states in the south-eastern corner of the country. It has several morphological characteristics not shared by other *Kalotermes* species, including a nearly straight posterior margin to the pronotum. There are 13 antennal segments, and segment 3 is nearly as long as segments 4 and 5 combined. This species is considered to be an uncommon structural pest.

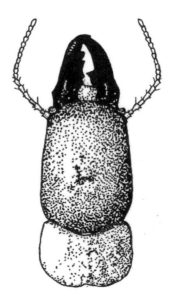

Fig 5.6
Head and pronotum of a
***Kalotermes flavicollis* soldier.**

Kalotermes flavicollis: A widespread, though generally minor, construction pest throughout the humid parts of the Mediterranean zone. It is found in the dead or diseased wood of a wide variety of trees, usually at the boundary with healthy wood. If sufficient moisture is present it will continue to thrive in dead trees and building timbers. Activity within the colony is seasonal, ceasing during the winter months. A mature colony may contain 1,000–1,500 individuals with only a small proportion of soldiers, and will take several years to develop (Fig 5.6).

Marginitermes

This termite soldier is distinctive because the third antennal segment is pigmented and as long as the next five to seven together. The mandibles are slender. There are two species in central and south-western America and a third in Australia.

Marginitermes hubbardi: Arid regions of America, Mexico and Central America, where it nests in dead trees and dead saquaro cacti. Urbanisation has allowed it to spread to construction timbers and it is now a major structural pest in south-western deserts in the USA (Fig 5.7). *M. cactiphagus* replaces it in the milder arid regions of Mexico and Central America.

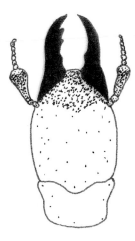

Fig 5.7
Head and pronotum of a
***Marginitermes hubbardi* soldier.**

5.4.3 Family: Hodotermitidae

These are known as harvester termites. All castes have functional compound eyes, while only rudimentary ocelli may or may not be present. The soldiers occur in two sizes and slightly differing forms (dimorphic) and their heads are large and thick. The antennae are long with 22 to 33 segments. The pronotum is saddle shaped with lateral elongations, and mandibles have large and pointed teeth on their inner margin. There are three marginal teeth on the left mandible and two on the right. The cerci have three to eight segments.

These are grass feeders, and their functional eyes allow them to forage in the open from late afternoon to early morning, or all day if they have constructed a mud sheet over their food resource. They nest in a series of interconnected subterranean hives. There are about 20 species in three extant genera.

Anacanthotermes

These termites have rather slender toothed mandibles and prominent eyes. They are found from Algeria to north-east India. They require semi-desert conditions where the soil has some clay content and there is enough water to support a sparse vegetation. Their grass-feeding habits can cause buildings to collapse if they remove straw from mud bricks, and they will also consume palm thatch and softer building timbers.

Anacanthotermes ochraceus: Distributed from Algeria to Libya, Egypt, Sudan and Saudi Arabia. Nests are constructed about 30cm below the ground and may extend down to around 1.5m. Each nest consists of a series of interconnecting round hives.

It has become an increasingly important building pest species as land has been irrigated for agriculture and many different timbers and wood products are imported (Fig 5.8).

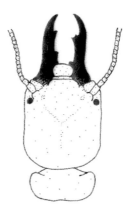

Fig 5.8
Head and pronotum of an
Anacanthotermes ochraceus
soldier.

Other potential pest species which differ in colour or slight physical characteristics but are similar in habit include:

Anacanthotermes septemtrionalis: Iran, Afghanistan, Turkmenistan.

Anacanthotermes ubachi: Southern Israel, Jordan and Iraq. Soldiers are dark brown and the sides of heads are slightly concave.

Anacanthotermes vagans: Iran, Iraq, Afghanistan and Pakistan. Pale yellow and sides of head straight.

Hodotermes

Hodotermes mossambicus: South and East Africa. The main diet is dead grass, but plant litter is also consumed, including building thatch and timber (Fig 5.9).

Fig 5.9
Head and pronotum of a
Hodotermes mossambicus
soldier.

5.4.4 Family: Rhinotermitidae

These are the most advanced of the lower termites, with symbiotic protozoans in their hind gut. A fontanelle – an opening of the frontal gland – is present, but variable in size and placement. In soldiers it becomes a defensive organ emitting a sticky fluid. The antennae have 12–17 segments, the pronotum is flat, and the cerci are reduced to two segments. The mandibles lack prominent teeth.

These are subterranean nesters, but with the ability to form colonies above ground if there is a suitable supply of moisture – perhaps from a drainage fault. There are about 168 living species, in about 13 genera. These are organised into six subfamilies, of which four contain species which are of significant concern in relation to heritage protection.

Subfamily: Coptotermitinae

This is the most primitive of the subfamilies and contains only one genus, *Coptotermes*, and 23 species which are distributed in all tropical and subtropical regions. They cause extensive damage to trees and invade buildings and other man-made structures. They are mostly very similar in appearance and there is considerable taxonomic confusion (Chouvenc *et al* 2016).

The head capsule is oval and the fontanelle is well developed, with a large prominent opening on or near the clypeus (Plate 7). The mandibles are slender and curved, without teeth but often with crenulated bases.

Coptotermes

Sixteen species are mentioned here to show the range of distribution and the ease with which these termites have been spread around the world. This list is not intended to be complete and other *Coptotermes* species are also arboriculture pests and may damage timbers in buildings.

Coptotermes acinaciformis: Australia, introduced into New Zealand. This termite usually nests at the base of a tree trunk, but may be subterranean if there is buried timber debris, and may form a mound. Living trees are invaded by boring out the core and filling the cavity with frass. The foraging distance from the nest may be 50m over perhaps 0.5

acres. Subsidiary nests may be formed in building cavities if moisture will allow, and a colony may contain 1,000,000 individuals. This species has an extensive distribution in Australia and causes considerable damage to trees and building timbers.

Coptotermes amanii: Somalia, Kenya, Tanzania, Malawi, Zambia and Zimbabwe. This species causes considerable damage to buildings in coastal Tanzania and Kenya.

Coptotermes ceylonicus: India and Sri Lanka, where it causes significant damage to buildings.

Coptotermes crassus: Mexico, Honduras and Guatamala.

Coptotermes formosanus: Taiwan, China, Japan, Pacific Islands, Sri Lanka, South Africa and USA. Now found across the Indian subcontinent. In the 1950s it was reported from South Africa. It was recorded in Hawaii in 1913, from a shipyard warehouse in Houston, Texas, in 1965, and has now spread extensively. The termite requires a high humidity and is potentially invasive in a zone which extends 35° north and south of the equator. A colony may contain over 1,000 termites in 3 years, 50,000 in 4 years, and continue for more than 50 years, when there will be several million individuals. A swarm may then contain 70,000 alates. These termites will burrow down to about 3m and a colony may forage over 0.5 acres of land (Fig 5.10 and Plate 7).

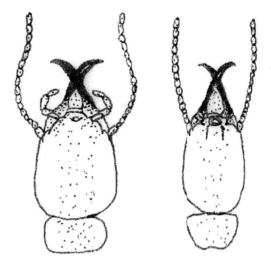

Fig 5.10
Heads and pronotums of
Coptotermes formosanus (left)
and *C. frenchi* soldiers.

Coptotermes frenchi: Australia (*see* Fig 5.10).

Coptotermes gaurii: Sri Lanka.

Coptotermes gestroi: Malaysia, Thailand and India. It is the most important subterranean termite pest in Thailand. It is an introduced species in

Madagascar, Mauritius, Barbados, Jamaica and many other West Indian islands. It was reported from Brazil in 1923, probably imported in marine cargo, and has become the most important urban termite pest. Soldiers are aggressive and will rapidly move to the site of a disturbance. This species was also known as *Coptotermes havilandi*.

Coptotermes grandiceps: Solomon Islands, where it is found in construction timbers.

Coptotermes heimi: Bangladesh, Bhutan, India and Pakistan, where it is found in construction timbers, joinery and books.

Coptotermes niger: Bahamas, Belize, Guatemala, Panama and West Indies, where it is a pest in urban buildings.

Coptotermes premrasmii: Thailand, where it is a pest in rural and urban buildings.

Coptotermes sjostedti: West Africa (Senegal to Cameroon), Congo, Angola and Uganda. A rainforest species which causes extensive damage to buildings along the coast.

Coptotermes testaceus: Bahamas, West Indies, Venezuela, Guianas, Trinidad, Brazil, Chile, Peru, Bolivia, Brazil, South and Central America. Natural nest sites are damp wood in forests where it may be a pest in managed forests, but it also causes considerable damage to buildings.

Coptotermes travians: Malaysia, where it attacks ships and wood that is damp because of building or plumbing faults.

Coptotermes truncatus: Madagascar and Seychelles, where it is found in building timbers.

Coptotermes vastator: Philippines, where it is highly destructive to woodwork in buildings. Introduced to Hawaii.

Subfamily: Heterotermitinae

This subfamily contains two genera. *Heterotermes* has about 36 species which are found in all tropical regions, while *Reticulitermes* is a temperate genus with about 14 described species. *Heterotermes* seems better adapted to arid conditions and will even attack driftwood on hot and dry beaches. The two genera are mostly geographically isolated.

The soldiers' heads have parallel sides and the antennal segments are reduced in number. The fontanelle is circular and minute. The labrum is longer than broad. That of *Heterotermes* tends to be needle shaped or pointed, while the labrum of *Reticulitermes* tends to be blunt. The mandibles of *Heterotermes* are thin and slightly curved at the tip so that they cross when fully closed. *Reticulitermes* mandibles are thicker and do not cross. There are 12–19 antennal segments.

The alates of *Heterotermes* are pale yellow, while those of *Reticulitermes* are dark brown.

Heterotermes

Heterotermes aureus (desert subterranean termite): Mexico and USA in the desert regions of California and Arizona. This is the most destructive termite in Arizona, damaging homes and other man-made structures. It can survive both heat and desiccation and is therefore found in the hottest and driest areas. Jones (1990) calculated that a population might contain 300,000 foragers and that the average foraging party she investigated was 1,456 individuals, of which 8.6% were soldiers (Fig 5.11).

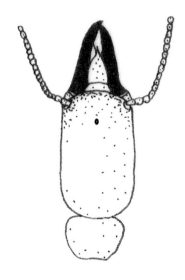

Fig 5.11
Head and pronotum of a
Heterotermes aureus soldier.

Heterotermes ceylonicus: India and Sri Lanka, where it attacks woodwork in houses.

Heterotermes convexinotatus: Central America, Venezuela, Colombia and West Indies. Once considered to be mostly a crop pest, this species is becoming an important structural pest in Colombia. Damage is indicated by thin foraging mud tunnels that pass through walls and ceilings and are often only noticed when the termite population has built up. Buildings become particularly at risk of collapse in areas where there is seismic activity.

Heterotermes gertrudae: India, where it is a building pest, particularly in the western Himalayas.

Heterotermes indicola: Indian subcontinent to Pakistan and Afghanistan. It is a common and serious pest in building timbers, where it attacks anything containing cellulose.

Heterotermes longiceps: Brazil and Argentina.

Heterotermes perfidus: St Helena. It was introduced into the island during the 19th century from an unknown source and is now a significant building pest.

Heterotermes philippinensis: Philippines, introduced into Madagascar and Mauritius. It is a significant pest in buildings.

Heterotermes tenuis: Central America, South America and the West Indies. It attacks anything containing cellulose in buildings.

Reticulitermes

These termites have a northern temperate distribution (Palearctic and Nearctic) and have recently been introduced into South America. Populations show seasonal changes in feeding behaviour and migrate deeper into the soil in cold weather. It is the only termite genus found where severe winters may occur. They are found in every continental state in the USA except Alaska. There are currently seven European species and perhaps there should be more, but the taxonomy of the genus is so difficult that there is no consensus of opinion as to how they should be identified.

Reticulitermes chinensis: China and Japan, where it damages buildings, feeding on wood and vegetable fibre. It also attacks the heartwood of old living trees.

Reticulitermes flavipes (eastern subterranean termite): Entire eastern region of North America from Canada to Florida. It is a significant pest species that may have been introduced into Austria and Germany, although the taxonomy of the genus is too complex to provide a definitive opinion (Fig 5.12 and Plate 8).

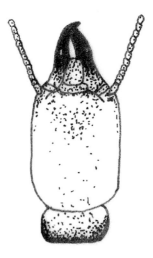

Fig 5.12
Head and pronotum of a
Reticulitermes flavipes soldier.

Reticulitermes grassei: Spain, Portugal and south-west France, introduced into the UK. Previously thought to be a subspecies of *R. lucifugus*.

Reticulitermes hageni (light south-eastern subterranean termite): This termite is found in Florida.

Reticulitermes hesperus (western subterranean termite): USA, along the Pacific coast from British Columbia to California.

Reticulitermes lucifugus: Mediterranean countries, Israel, Turkey. Distribution is by alates.

Reticulitermes santonensis: France, particularly in urban areas and a contender for the termite in Germany (Hamburg). Possibly introduced into Chile. This species has the most northern distribution of the seven species found in Europe. It is thought to have been distributed by the movement of rail freight and to have been encouraged to colonise buildings by central heating. Distribution is mainly by secondary reproductives. It may be the same species as *R. flavipes* (see above).

Reticulitermes speratus: China, Japan, Korea and Taiwan, where it is an important pest species.

Reticulitermes tibialis: USA in the western inter-mountain region, where it favours open, dry and sunny locations. Its range extends to Mexico.

Reticulitermes virginicus (dark southern subterranean termite): USA, southern states.

Subfamily: Prorhinotermitinae
Soldiers are of two different sizes, with slight differences in shape (dimorphic) and their heads are rounded with a small fontanelle. The antennae have 15–19 segments. These termites nest in dead trees, logs or stumps and decaying damp wood. Their oceanic distribution suggests that infested logs may have been transported by ocean currents.

Prorhinotermes
The genus contains about 20 species, mostly in Southeast Asia. The head shape and mandibles of the soldiers resemble those of *Coptotermes*, but the fontanelle is small and on top of the head, while that of *Coptotermes* is large and points forwards above the mouthparts.

Prorhinotermes simplex (Cuban subterranean termite): USA (south-eastern Florida), Puerto Rico and Jamaica. It is found in subtropical forests, mangroves and urban environments along the coast and on islands. Structural damage is usually confined to timbers in contact with the ground (Fig 5.13).

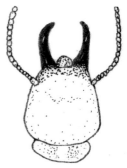

Fig 5.13

Head and pronotum of

***Prorhinotermes simplex* soldier.**

Subfamily: Psammotermitinae

This subfamily contains two genera. Soldiers occur in two or three different sizes (di- or trimorphic). The head is rectangular with a small fontanelle. Mandibles have five to nine marginal teeth. The antennae have 12–19 segments and the pronotum is flat.

Psammotermes

Psammotermes contains four species from arid regions and is of significance in relation to heritage protection. Species are easily separated by region.

Psammotermes allocerus: South and West Africa, where it causes considerable damage to construction timber, joinery and books.

Psammotermes hybostoma: Desert regions of North Africa and the Arabian Peninsula. The pronotum is densely hairy (Fig 5.14).

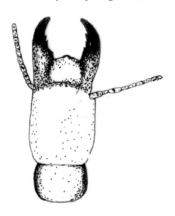

Fig 5.14
Head and pronotum of a
Psammotermes hybostoma
soldier.

Psammotermes rajasthanicus: North-west India and Pakistan, where it damages woodwork in buildings.

Psammotermes voeltzkow: South-west Madagascar, where it is a significant pest in building timbers.

Subfamily: Rhinotermitinae

Each mandible has a few sharp pointed teeth. The fontanelle is small and at about the level of the antennae with a forward-running groove. Some species are dimorphic.

Schedorhinotermes

The head is nearly as wide as it is long, with sickle-shaped mandibles with sharp marginal teeth. The soldiers are dimorphic. The labrum is broad with a bifurcated tip, and in minor soldiers it may reach almost to the end of the mandibles. There are 16–17 antennal segments. This genus mostly lives in decaying timber, such as tree stumps, the root crowns of dead or dying trees or buried timber. They are able to tolerate saline conditions and so may be transported by sea.

Schedorhinotermes intermedius: Australia. This is a significant pest in wooden buildings around Sydney (Fig 5.15).

Fig 5.15
Head and pronotum of a
Schedorhinotermes intermedius
soldier.

Schedorhinotermes lamanianus: Tanzania and South Africa. This is a humid lowland forest species.

Schedorhinotermes medioobscurus: Thailand and Malaysia, where it is a pest in urban and rural buildings.

Schedorhinotermes solomonensis: Solomon Islands.

5.4.5 Family: Termitidae

The Termitidae are considered to be higher termites and lack protozoa in their hind gut. They include the majority of the described termite species. Soldiers are very diverse in form. Genera may have one, two or three sizes and forms of soldiers (monomorphic, dimorphic or trimorphic). There are 11–20 antennal segments. The pronotum is saddle shaped, narrower than the head, and the anterior margin is lobed. There are seven subfamilies, of which three contain species of particular concern in relation to heritage protection.

Subfamily: Macrotermitinae

African and Southeast Asian wood feeders that construct complex subterranean or epigeal nests with fungus combs. Soldiers have sabre-shaped mandibles that are either smooth or have a single marginal tooth.

Macrotermes

Soldiers' mandibles do not have teeth.

This genus, which produces large and conspicuous mounds, includes some of the biggest termites. A study of rural housing in Ethiopia (Debelo and Degaga 2014) found that 79% of the termites causing destruction belonged to the genus *Macrotermes*. These houses had mostly mud walls with wood and straw thatch roofs. They were destroyed in five to six years. The African species listed below are recorded as serious pests in buildings (Edwards and Mill 1986).

Macrotermes bellicosus: West Africa.

Macrotermes gilvus: Malaysia, where mounds are found under floors and the termites forage up through the building timbers. This problem is particularly common where plantation land is cleared for human settlement (*see* Fig 2.8) and timber debris has been left (Fig 5.16).

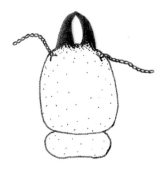

Fig 5.16
Head and pronotum of a
Macrotermes gilvus soldier.

Macrotermes natalensis: Central Africa and South Africa.

Macrotermes subhyalinus: East, West and Central Africa and Sudan.

Odontotermes

This is a large genus of fungus-growing termites that contains around 200 species, distributed in the Ethiopian, Oriental Palaearctic and Papuan regions. Several cause considerable damage to building timbers. These are subterranean termites that sometimes form mounds. Colonies may be very large, producing nests that are more than 1m in diameter (Fig. 2.9). The mandibles of the soldiers are large and strong. The left mandible has one lateral tooth and the position of this tooth may be the only morphological difference between species.

Species of the small Asian genus *Hypotermes* are very similar in appearance and habits to *Odontotermes* except that they lack a tooth on their mandibles.

Odontotermes badius: Tropical Africa, Namibia and South Africa. A pest of groundnuts, sugar cane and other crops, that also attacks building timbers. The defensive secretions of this species have been studied and soldiers produce an opaque brown fluid that hardens into a rubbery and sticky tangle on exposure to air (Wood *et al* 1975).

Odontotermes ceylonicus: Sri Lanka. This species does not produce mounds. It lives on the roots of dead and dying trees and forages into buildings, where it is a structural pest. It is frequently found living in the nests of other termites.

Odontotermes distans: India, Bangladesh, Bhutan and Myanmar. This species feeds on bark, wood debris, dead leaves, cow dung and sometimes the sapwood of living trees (Thakur 1981). It attacks building timbers if they are in the vicinity of the nest.

Odontotermes feae: Bangladesh, Myanmar and India. It attacks all cellulose-based materials and is one of the commonest building termites in India (Fig 5.17).

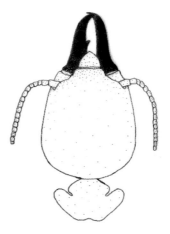

Fig 5.17
Head and pronotum of an
Odontotermes feae **soldier.**

Odontotermes formosanus: Taiwan, Vietnam, Cambodia, Thailand, China and Japan. A serious pest of forests, crops and buildings that weakens dams and dykes.

Odontotermes horni: Sri Lanka. It attacks bark and dead wood and is a significant pest in tea and rubber plantations.

Odontotermes javanicus: Malaysia and Indonesia, where it seems to be an emerging pest as plantation and forest land is used for expanding urban housing.

Odontotermes obesus: India and Pakistan. A serious pest in plantations, forests and buildings.

Odontotermes obscuriceps: Sri Lanka. This species attacks the stems of tea plants and, like other *Odontotermes*, any building timbers that are in the vicinity of the nest.

Odontotermes pauperans: Nigeria to Senegal. A mound-building species that influences the pattern of growth of some savannah grasses.

Microtermes
A genus with more than 100 species distributed in most tropical regions. Colonies are moderate in size and some species construct small mounds. These termites live on dead plant material and may forage within plant

stems. Soldiers are around 5mm long and have elongated rectangular heads and mandibles with incurved tips. They are short, thin and delicate. The soldiers are smaller than the workers.

Microtermes obesi: India, Pakistan, Sri Lanka and Vietnam. A pest of wheat and sugar cane that forages into building timbers (Fig 5.18).

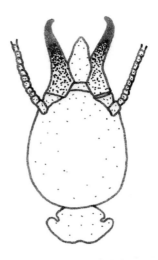

Fig 5.18
Head and pronotum of a
Microtermes obesi soldier.

Microtermes pakistanicus: Thailand. A pest of rubber and oil palm plantations that invades building timber.

Microtermes redenianus: Tanzania. Another significant crop pest that forages into buildings.

Subfamily: *Nasutitermitinae*
These are tropical termites with no visible mandibles and the front of the head projected into a snout (nasus). Many construct arboreal nests, some are subterranean, and a few construct mounds. They mostly feed on lichen or dead leaves and branches.

Nasutitermes
Soldiers have a nasus, an elongated projection of the head from which they can squirt a defensive secretion. This effective secretion has even been shown to deter Tamandua ant eaters (Lubin and Montgomery 1981). Mandibles are considerably reduced in size. Nests are commonly arboreal but may also be mounds or subterranean. *Nasutitermes* feed on dead wood in living trees. They construct an extensive foraging tube system to connect nesting and feeding sites. and readily attack building timbers. They sometimes seem to nest as a colony and there may be several primary and secondary reproductives present (Atkinson and Adams 1997). Colonies may contain over a million termites.

Nasutitermes ceylonicus: Sri Lanka. A pest of coconut and tea that forages into buildings.

Nasutitermes corniger (conehead termite): Guyana, Guatemala, Panama, West Indies, Puerto Rico and Florida. Now introduced to Papua New Guinea. These are small termites with a soldier length of about 5mm. Alates have black wings and dark bodies. This is an aggressive pest that will attack anything containing cellulose (Fig 5.19).

Fig 5.19
Head and pronotum of a
Nasutitermes corniger soldier.

Nasutitermes ephratae: Guadeloupe, Guatemala, Montserrat, Panama and Trinidad. This species is found over the same range as *N. corniger*. Both produce arboreal nests in lowland habitats. The soldiers of the two species are difficult to differentiate, but the alates of this species have yellow/brown wings and brown bodies. The ocelli are also located close to the eyes.

Nasutitermes exitiosus: South and West Australia. Common in Canberra and New South Wales. Eisner *et al* (1976) studied defence in this species and found that the workers also attacked predators, slowing them down so that they were sprayed by the soldiers. The secretion was a viscous entangling irritant.

Nasutitermes globiceps: Paraguay, Bolivia and Brazil. This is a significant pest species causing damage to buildings, transmission posts, fences, trees and stored materials.

Nasutitermes javanicus: Thailand and Malaysia. A pest of oil palm plantations that invades building timbers.

Nasutitermes voeltzkowi: Mauritius.

Subfamily: Termitinae
Pantropical distribution with species feeding on the full range from sound to totally decomposed plant materials.

Amitermes
This is a large tropical and subtropical genus. They usually nest in the soil and most produce mounds, occasionally of considerable size. Foraging tubes are constructed to connect nests and feeding sites above and below ground. The soldiers have round or pear-shaped heads with sickle-shaped mandibles, each with a sharp-pointed median tooth. Antennae have 13–17 segments.

Amitermes herbertensis: Australia. A rain forest insect that lives in rotten logs and tree stumps. Destroys poles and other structural timbers in contact with the soil.

Amitermes messinae: Africa, Saudi Arabia to South Africa. A subterranean termite forming dark carton nests incorporating timber. A pest within the limestone walls of buildings in Bahrain (Fig 5.20).

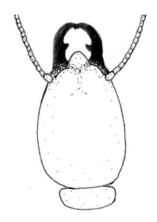

Fig 5.20
Head and pronotum of an
Amitermes messinae soldier.

Amitermes vilis: Iran and Turkmenistan. Commonly nests within dead tree stumps. This is one of the most economically important termites in Iran, where it causes damage to crops and monuments in agricultural and urban areas.

Amitermes wheeleri: USA, where it is widespread and common. It damages woodwork in buildings.

Globitermes
A genus of small-mound termites from Southeast Asia with few species. Soldiers are similar to *Amitermes*, but have rounder heads.

Globitermes sulphureus: Thailand, Malaysia and Vietnam. This species is a significant pest in coconut and oil palm plantations, but also attacks timbers around the perimeters of buildings (Fig 5.21). Soldiers have yellow abdomens.

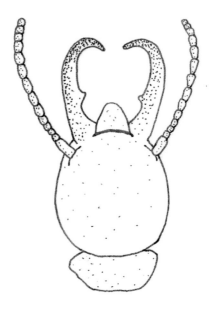

Fig 5.21
Head and pronotum of a
Globitermes sulphureus soldier.

Microcerotermes

This genus occurs throughout the tropics. The soldiers have long, dark-brown heads and long mandibles with incurved tips. The inner margins may be serrated when viewed with a microscope. Antennae have 12–15 segments. Nests may be subterranean, small mounds, arboreal and sometimes on the tops of posts.

Microcerotermes crassus: Thailand and Malaysia (Fig 5.22).

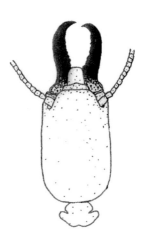

Fig 5.22
Head and pronotum of a
Microcerotermes crassus soldier.

Microcerotermes diversus: Israel, Iran, Iraq and Oman. A major pest of date palms that causes considerable damage to building timbers. Nests are initially diffuse and subterranean, but others are formed in trees or within walls.

Microcerotermes fuscotibialis: Sierra Leone. This species is a pest of cocoa farms in West Africa that produces earth foraging tubes lined with faecal material that cover tree trunks and earth or timber constructions. The termites cause significant damage to buildings.

References

Atkinson, L and Adams, E S 1997 'The origins and relatedness of multiple reproductives in colonies of the termite *Nasutitermes corniger*'. *Proc R Soc Lond Series B* **264**, 1131–6, https://doi.org/10.1098/rspb.1997.0156

Buczkowski, G and Bertelsmeier, C 2017 'Invasive termites in a changing environment: A global perspective'. *Ecol Evol* **7** (3), 974–85

Chouvenc, T, Helmick, E E and Su, N-Y 2015 'Hybridization of two major termite invaders as a consequence of human activity'. *PLoS ONE* **10** (3), e120745, https://doi.org/10.1371/journal.pone.0120745

Chouvenc, T, Li, H-F, Austin, J, Bordereau, C, Bourguignon, T, Cameron, S, Cancello, E and Constantino, R 2016 'Revisiting *Coptotermes* (Isoptera: Rhinotermitidae): A global taxonomic road map of species validity and distribution of an economically important termite genus'. *Syst Entomol* **42** (2), 299–306

Debelo, D G and Degaga, D G 2014 'Preliminary studies of termite damage on rural houses in the Central Rift Valley of Ethiopia'. *Afr J Agric Res* **9** (39), 2901–10

Edwards, R and Mill, A E 1986 *Termites in Buildings: Their Biology and Control*. East Grinstead: Rentokil

Eisner, T, Kriston, I and Aneshansley, D J 1976 'Defensive behaviour of a termite (*Nasutitermes exitiosus*)'. *Behav Ecol Sociobiol* **1**, 83–125, https://doi.org/10.1007/BF00299954

Evans, T A, Forschler, B T and Grace, J K 2013 'Biology of invasive termites: A world wide review'. *Annu Rev Entomol* **58**, 455–74

Grace, J K 1992 'Termite distribution, colony size, and potential for damage' *in* Robinson, W H (ed), *Proceedings of the National Conference on Urban Entomology*. College Park, Maryland, 23–26 February 1992, 67–76

Grace, J K, Woodrow, R J and Oshiro, R J 2009 'Gallery systems of one piece termites (Isoptera: Kalotermitidae)'. *Sociobiol* **54** (1), 37–44

Gulmahamad, H 1997 'Naturally occurring infestations of drywood termites in books'. *Pan-Pac Entomol* **73** (4), 245–7

Harris, W V 1961 *Termites: Their Recognition and Control*. London: Longmans

Harris, W V 1970 'Termites of the Palearctic region' *in* Krishna, K and Weesner, F (eds) *Biology of Termites*. New York and London: Academic Press, vol 2, 295–313

Harvey, P A 1934 'Biology of the dry-wood termite' *in* Kofoid, C A, Light, S E, Horner, A C, Randall, M, Herms, W B and Bowe, E E (eds) *Termites and Termite Control*. Berkeley: University of California Press, 217–33

Jones, S C 1990 'Colony size of the desert subterranean termite *Heterotermes aureus* (Isoptera, Rhinotermitidae)'. *Southwest Nat* **35** (3), 285–91

Krishna, K, Grimaldi, D A, Krishna, V and Engel, M S 2013 'Treatise on the Isoptera of the world'. *Bull Am Mus Nat Hist* **377** (1)

Lubin, Y D and Montgomery, G G 1981 'Defenses of *Nasutitermes* termites (Isoptera, Termitidae) against Tamandua anteaters (Edentata, Myrmecophagidae)'. *Biotropica* **13** (1), 66–76, https://doi.org/10.2307/2387872

Sheffrahn, R 2019 'Expanded New World distribution in the termite family Kalotermitidae'. *Sociobiol* **66** (1), 136–53

Sheffrahn, R and Krecek, J 1999 'Termites of the genus *Cryptotermes* Banks (Isoptera: Kalotermitidae) from the West Indies'. *Insecta Mundi* **13** (3–4), 111–71

Thakur, M 1981 'The identity, distribution and bioecology of *Odontotermes distans* Holmgren et Holmgren (Isoptera: Termitidae: Macrotermitinae)'. *Proc Indian Acad Sci* **90** (2), 187–93

Wood, W, Truckenbrodt, W and Meinwald, J 1975 'Chemistry of the defensive secretion from the African termite *Odontotermes badius*'. *Ann Entomol Soc Am* **68** (2), 359–60

6 Termite detection

Inspections for decay and insect infestation in buildings are invariably restricted by accessibility. An inspection can never be complete unless the building fabric can be opened up to reveal what lies within. Even in small domestic buildings there will usually be hidden construction features, including voids and cavities, within walls and floors. Understanding the construction and condition of heritage buildings can be further complicated by past repairs, alterations and the effects of ageing. Evidence will often be hidden behind important surface finishes, so the ability to carry out investigations will be constrained by the need to avoid harming fabric of heritage significance. The aim of conservation is to conserve and this is not achieved if the inspection causes more damage than the organisms that need controlling.

6.1 Visual detection

6.1.1 Drywood termites
The termite colony lives within a single piece of wood, which may be brought into a building, or swarming alates may fly in attracted by light. Feeding is therefore confined and the colony may die out when accessible and suitable wood has been consumed. Wide feeding galleries within the timber are connected by narrow access passageways, which are easily blocked to hinder predators.

Fig 6.1
Kalotermes flavicollis pellets below a 'kick hole' in the base of a door frame (Skopelos, Greece).

Periodically they make a small hole to the exterior and eject the accumulated faecal pellets (Fig 6.1). These pellets may be the first indication that the termites are present. They are hard, oval and about 1mm long, with rounded ends and six longitudinal facets. Their colour is quite bright when fresh but becomes dull with age.

Nearly all wood that is not treated or naturally resistant will eventually be consumed, but a thin veneer of timber (or perhaps just paint) will remain at the surface for protection and to retain humidity (Fig 6.2). The collapse of structures such as floorboards may be the first indication of a termite attack.

Fig 6.2
Cryptotermes brevis frass pellets under a thin veneer of timber on a floorboard containing a nest (Grand Turk Island, Turks and Caicos Islands).

Alates may swarm within the building and appear on windows. There might also be piles of shed wings (*see* Plate 5).

Sutherland *et al* (2014) experimented with a borescope to aid inspection of cavities and inaccessible spaces. Borescopes (or endoscopes) have an integral light source and a fibre-optic light guide with an objective lens at one end and an eyepiece or display at the other. A borescope may be long, short, rigid or flexible. The experimental procedure used by Sutherland was to provide a range of samples, including heaps of frass and dead termites, in a laboratory and a test building. These were concealed from view and seven surveyors with varying amounts of experience were asked to identify them. The authors reported a mean detection accuracy of 80.6%.

Borescopes have been used by this author over the last 40 years in the UK to detect dry rot and wood-boring beetle holes. The main limitation is the effective working distance – it can be difficult to discern a 1mm-diameter hole when it is 500mm away from the end of the scope. Systems commonly used require 15mm-diameter inspection holes. Narrower borescopes are available, but light levels are reduced. Borescopes can be inserted through existing apertures in floorboards and wall linings. This makes them a useful tool in situations where opening up is not acceptable or permitted. However, they have limitations and important information may be missed when cavities are large or filled with debris.

6.1.2 Subterranean termites

It is commonly assumed that colony initiation within a building by subterranean termite alates would be unusual and that most infestations commence by worker termites foraging up from the soil. This, however, is not necessarily true for historic structures, which may be complex or in poor condition, perhaps with earthen floors and courtyards. Case Study 1 provided a probable example, but an unequivocal demonstration was supplied by the Golden Temple at Amritsar.

Termite tubes began to appear every morning between the gilded copper sheets in the ceiling of the entrance lobby. These contained *Coptotermes sp*. termites which were foraging down from a colony established in blockboard shelving in a storeroom above. The termites must either have been introduced in the shelving, which was unlikely because it had been cut to size, or by alates following a swarm. This seemed the most probable explanation because the temple stands on a stone block in a huge artificial lake. The source of moisture was frequent roof washing, probably combined with drainage that had become faulty over the centuries. However, a more detailed inspection was neither possible nor necessary. The shelving could be removed and treatment was not required because there was no wood in the construction.

In most cases, foraging subterranean termite workers will enter buildings through cracks in the foundations where they can pass directly from the soil without exposing themselves to light and the open air. They then proceed from the ground via foraging tubes on walls and in subfloors. These are the most frequently observed sign of termites. They may be obvious on surfaces of a contrasting colour, but more difficult to find between wooden elements or on dirt-streaked foundations. They are an extension of the nest and mound construction and enable the workers to locate and exploit food resources above ground by protecting them from the environment and predators (particularly ants). Tube-building behaviour is controlled by a 'cement pheromone' (Mizumoto *et al* 2015). This pheromone is secreted and infused into soil particles when construction commences. It recruits more workers, which accumulate more pheromone, increasing the rate of building until the construction – mound, nest wall or foraging tube – is complete. Construction and repair can thus proceed fairly rapidly, and one investigation observed that a 3cm-long broken section of tube was completely restored in 65 minutes (Exner 1953).

Foraging tubes vary in size and material according to termite genus. Subterranean nesting wood feeders, like most *Coptotermes*, line their galleries (Fig 6.3) and construct their tubes from carton mixed with soil. Macrotermitinae genera such as *Odontotermes* may use more clay and sand moistened with saliva. If tube material is in small gaps, perhaps between boards (Fig 6.4), it may require a close inspection to find. Waving a torch from an attic hatch will not be sufficient.

Fig 6.3
Reticulitermes sp. galleries
lined with tube material.

Fig 6.4
Odontotermes sp. tube
material between ceiling
boards (Ethiopia).

Four categories of tubes may be recognised:

Exploratory tubes: These foraging tubes are generally thin walled and multi-branching to locate suitable food resources. They will rarely have termites in them and they are often found as broken remnants.

Working tubes: These are foraging tubes formed when a food resource has been discovered and have thicker walls that are kept in good repair while the resource is exploited. Tubes are generally around 5mm–10mm in diameter, wide enough to allow two streams of termites to pass – one going to the food source, the other returning to the nest. Breaking an active tube will expose workers, perhaps soon joined by soldiers for protection while the damage is repaired. Some of the higher termites expand the tube material into sheets that cover exposed wood.

Drop tubes: Drop tubes are sometimes suspended between the food resource and the ground and are formed just from the food material. Their function seems to be to convey extra moisture to the feeding sites.

Swarming tubes: Rhinotermitidae construct tunnels and tubes to conduct the alates to the most advantageous location for dispersal. This may be the side of the building where night temperatures are most favourable. The Termitidae also make elaborate preparations, particularly the Macrotermitinae, which produce wide shafts with the apertures sloped and free from debris to form launching platforms. Swarm tubes may be wide to allow alates to congregate prior to dispersal.

In timber, the termites usually commence by burrowing along the grain, particularly if there is soft early wood, as found in pine or teak. The presence of damage can often be detected by a hollow sound when the wood is struck, or the surface may collapse. The termites may reinforce the damage by constructing walls or, if they are subterranean species (particularly Termitidae), filling in gaps with chewed wood, often mixed with soil or sand.

6.2 Detection devices

Visual inspection has limitations because the signs of an infestation might be in inaccessible places, and piles of wings do not necessarily direct attention to the infested timber. An infestation may be historic and inactive, or if it is active, it might not become apparent until significant damage has been caused.

The simplest commercial device to use would be a moisture meter. Dry or damp readings, at least with subterranean termites, may help to distinguish between historic and current damage. Use as a survey instrument may also locate damp areas that require further investigation and maintenance work. It should be remembered, however, that the moisture meter is not measuring moisture. It is measuring some electrical property of the timber, which, in historic buildings, may have been influenced by factors such as timber age, surface coating or contamination from airborne or waterborne pollutants. If a meter reading indicates that a timber is dry (moisture content <15%) then it probably is, but higher readings should be treated with caution. A pattern of readings, particularly if this can include original together with modern or replacement components, may be needed to obtain a correct interpretation of the meter readings. The limitations and interpretations of meter readings are discussed in Ridout and McCaig (2023).

The difficulty in locating termites has led to the development of a wide range of detection devices with varying levels of efficacy, although all probably work in some situations. Although they have been shown to be effective in laboratory evaluations, buildings present a more complex challenge due to wide variations in construction and thickness of materials and finishes. Termite behaviour can also be a problem. Termites may be more active at different times of the day depending on diurnal temperature fluctuations, and activities may vary depending on the season. For example, searching for drywood termite activity in a cold structure during winter months may provide limited results because the termites are not active, rather than not present.

6.2.1 Detecting movement – microwave signals

Microwaves comprise the frequency range 0.3Ghz–300Ghz and lie between the radio and infrared regions of the electromagnetic spectrum. An Australian device, Termatrac™, emits a constant stream of microwaves, which are reflected back by moving objects. The manufacturers state that it is capable of detecting a single termite. This sensitivity means that the device might give a positive response to any kind of vibration, including a shaky hand and background movement. These problems can be overcome by carefully inspecting the environment, avoiding all movement during usage and perhaps using a tripod. If used carefully then the device can be very effective.

Evans (2002) found that it accurately detected activity in *Coptotermes lacteus* and *Nasutitermes exitiosus* in baited 'aggregation stations' – 117 steel boxes containing wooden slats that were either buried in the soil or placed on posts, depending on the termite. The Termatrac was placed on top of each box. A 90% success rate was obtained in the detection of presence or absence of termites. Some of the false readings were caused by other insects, but the instrument could distinguish between insect types if the readings were correctly interpreted. The success rate improved with experience. However, microwaves are absorbed by water, so the instrument may not be as effective on damp wood. In addition, signal attenuation with depth may give false negative results on thick timber. Peters and Creffield (2002) found that the device was effective up to a depth of 35mm in untreated pine boards and 25mm in treated pine and heartwood.

6.2.2 Detecting sound – acoustic emission

The termites that concern us here live in wood and wood can absorb, produce and amplify sound signals, thereby making it one of the best materials for constructing musical instruments. These properties can be utilised for termite detection. The foraging and feeding of the large *Mastotermes* are easily heard, but the sound from the activities of most other termites requires amplification to detect.

In 1929, Emerson and Simpson reported on the use of a telephone transmitter to hear the movement of termites and the head-banging communication of the soldiers. Technical advances in transducers and amplifiers substantially improved the method and it was used to locate a wide array of insect pests, but it was highly susceptible to background noise.

The problem was overcome by optimising acoustic emission detection at higher frequencies so that background interference was eliminated. Termites break or chew off fragments of timber and this can be detected as acoustic emission, which is the elastic energy released when a surface undergoes deformation. Highly sensitive piezoelectric transducers can convert this surface displacement into an electrical signal. Lemaster *et al* (1997) concluded that a transducer frequency of 60kHz was the most useful. Above this, there was a problem with attenuation, and below this, background noise became increasingly significant.

El-Hadad (2017) made a thorough study of the method and found that the frequency could be considerably reduced if filters were used.

Devices are commercially available with surface and subsurface probes. However, wall coverings can impede sensor use and performance.

A test using a hand-held, battery-powered device to detect subterranean and drywood termites in small blocks of wood proved to be 98–100% effective (Scheffrahn *et al* 1993). They found that the termites could be detected up to 80cm along the grain from the device, but only 8cm across the grain (tangential direction).

6.2.3 Detecting gases and odours

A range of atmospheric trace gases have been found to be produced by termites (Khalil *et al* 1990) of which the most significant are methane (CH_4) and carbon dioxide (CO_2). These authors also found that concentrations of chloroform ($CHCL_3$) in nests of *Coptotermes lacteus* were 1,000 times greater than ambient levels. Some termites produce a distinct and characteristic odour. This is particularly noticeable in the nests of *Nasutitermes exitiosus*.

These emissions have suggested that gas or odour detecting may be a useful survey method and devices – 'electronic noses' – have been marketed. One device that detected methane was tested by Lewis *et al* (1997). This had three settings (2, 4 and 6ppm) and a beeping sound was produced when the set level was exceeded. One hundred pine blocks were randomly populated with 0, 5, 50 or 200 termites. The odour detector only achieved a 48% correct identification rate, irrespective of the number of termites in the blocks and was not thought to be particularly useful. However, 'electronic nose' technologies continue to develop and higher success rates have been reported (Wilson and Oberle 2011).

Trained dogs have been available in the USA for termite detection by odour since the mid-1970s. Lewis *et al* (1997), using the wooden blocks containing termites as mentioned earlier, obtained a correct identification rate of 81%. The dogs performed best when there were more than 50 termites in the block.

Brooks *et al* (2003) trained dogs to detect eastern subterranean termites (*Reticulitermes flavipes*) in plastic containers, and found that they were 96% accurate if 40 or more termites were present. Similar or better success rates were obtained using the same dogs with *R. virginicus*, *Coptotermes formosanus* and the drywood termites *Cryptotermes cavifrons* and *Incisitermes snyderi*.

Zahid *et al* (2012) noted that in Australia there was great difficulty in ensuring that imported timber and timber products were free from pests, including termites. Current practice at that time was for an inspector to check for signs of insect activity and, if any were found, the item was placed in quarantine to see if the infestation was active. This process was lengthy and expensive. They tested seven detection methods (including visual inspection) by collecting 120 branches of *Acacia* spp. from a forest in Sydney. Visual inspection suggested that 80 of the branches were infested with insects. A further 12 wooden blocks were inoculated with beetles or termites. At the end of the experiment, all samples were taken apart to confirm presence or absence of insects. The dog, used by its handler and trained to detect termites, had a 100% success rate on the branches, but could not detect termites in the blocks when there were only 5–10 insects present. Visual inspection by two trained quarantine inspectors was 35% successful.

Most termite colonies are going to contain more than 50 insects and so trained dogs would seem to be effective. It is important to note that the dog should be used by its handler, with appropriate work and rest periods, so that its responses are understood. Dogs have been used to detect fungi in UK buildings, but there has been a tendency to regard the animal as another piece of equipment to be used by anybody. This has resulted in confusion. This author remembers being asked to evaluate a report that had the cryptic conclusion 'Dog barked once in porch.'

6.2.4 Detecting heat – thermal imaging

Infrared radiation is an invisible part of the electromagnetic spectrum.

Infrared thermography ('thermal imaging') uses a type of digital camera to produce visual images of infrared radiation emitted or reflected by objects and surfaces. The digestive organisms in the termite gut produce heat, and a feeding colony should be visible as bright spots in a thermal image. The surrounding areas are cooler and would appear darker. Cameras can measure temperature to about 0.1°C and should therefore be able to register the termite activity. The method is non-destructive and has the potential to quickly survey a large area of wall or floor. However, it is of limited use if an infestation is deep within wood and only a few termites are present.

The interpretation of images can be difficult, as variations in surface temperature will be influenced by many factors unrelated to termites. These include variations in the emissivity of surfaces, anomalies in construction, and ambient environmental conditions. Thermal imaging cannot 'see' beneath the surface and is therefore best used in conjunction with other techniques. Ultimately, the usefulness of the outputs will depend on the sensitivity and resolution of the equipment, and the skill and experience of the surveyor in interpreting the results.

6.2.5 Detecting internal damage – X-rays and T-rays

These are penetrating rays that are nearest to ultraviolet rays in the electromagnetic spectrum. Unlike other detection techniques, devices operating in the X-ray part of the spectrum allow users to 'see beneath the skin' of an object. They have numerous commercial and military applications and have also been used to find insects in wood. Research and development in this area continues, and ever more sophisticated imaging technologies are available, including computational tomography (CT) scanning. Imaging technologies using terahertz radiation ('T-rays') have also been used experimentally for detecting insect damage (Krügener *et al* 2019). This band of penetrating radiation lies between the microwave and infrared parts of the electromagnetic spectrum. However, equipment exploiting these technologies is expensive and generally not suitable for site survey work.

References

Brooks, S E, Oi, F M and Koehler, P G 2003 'Ability of canine termite detectors to locate live termites and discriminate them from non-termite material'. *J Econ Entomol* **96** (4), 1259–66.

El-Hadad, A 2017 *Using Acoustic Emission Technique with Matlab® Analysis to Detect Termites in Timber-in-Service*. Thesis submitted for the Degree of Doctor of Philosophy, University of Melbourne

Emerson, A E and Simpson, R C 1929 'Apparatus for the detection of substratum communication among termites'. *Science* **69** (1799), 648–9

Evans, T M 2002 'Assessing efficacy of Termatrac™; A new microwave based technology for non-destructive detection of termites (Isoptera)'. *Sociobiol* **40** (3), 575–83

Exner, W F 1953 'How fast can termites mend broken tubes'. *Pest Control* **21** (10), 52

Khalil, M A, Rasmussen, R A, French, J R and Holt, J A 1990 'The influence of termites on atmospheric trace gases: CH_4, CO_2, $CHCL_3$, N_2O, CO, H_2, and light hydrocarbons'. *J Geophys Res* **95** (D4), 3619–34

Krügener, K, Stübling, E-M, Jachim, R, Kietz, B, Koch, M and Viöl, W 2019 'THz tomography for detecting damages on wood caused by insects'. *Appl Opt* **58** (22), 6063, https://doi.org/10.1364/AO.58.006063

Lemaster, R L, Beall, F C and Lewis, V R 1997 'Detection of termites with acoustic emission'. *For Prod J* **47** (2), 75–9

Lewis, V R, Fouche, C F and Lemaster, R L 1997 'Evaluation of dog-assisted searches and electronic odour detecting devices for detecting the western subterranean termite'. *For Prod J* **47** (10), 79–84

Mizumoto, N, Kobayashi, K and Matsuura, K 2015 'Emergence of intercolonial variation in termite shelter tube patterns and prediction of its underlying mechanism'. *R Soc Open Sci* **2** (11), 150360, https://doi.org/10.1098/rsos.150360

Peters, B C and Creffield, J W 2002 'Termatrac™ microwave technology for non-destructive detection of insect pests in timber'. Document IRG/WP/02. Stockholm, Sweden: The International Research Group on Wood Preservation

Ridout, B V and McCaig, I 2023 *Assessing Dampness in Historic Building: A Practical Guide to Investigation and Diagnosis*. Swindon: Historic England

Scheffrahn, R H, Robbins, W P, Busey, P, Su, N-Y and Mueller, R K 1993 'Evaluation of a novel, hand-held, acoustic emissions detector to monitor termites (Isoptera: Kalotermitidae, Rhinotermitidae) in wood'. *J Econ Entomol* **86** (6), 1720–9

Sutherland, A M, Tabuchi, R L, Moore, S and Lewis, V R 2014 'Borescope-aided inspection may be useful in some drywood termite detection situations'. *For Prod J* **64** (7/8), 304–9

Wilson, A D and Oberle, C S 2011 'Development of an electronic-nose technology for the rapid detection and discrimination of subterranean termites within wood in service'. *Phytopathology* **101**, S192

Zahid, I, Grgurinovic, C, Zaman, T, De Keyzer, R and Cayzer, L 2012 'Assessment of technologies and dogs for detecting insect pests in timber and forest products'. *Scand J For Res* **27** (5), 492–502, https://doi.org/10.1080/02827581.2012.657801

7 Treatments and termite management

7.1 Termiticides

The traumatic first half of the 20th century promoted a massive increase in the development of pesticides. These were persistent and inexpensive so there was no need to search for alternatives. Unfortunately, their persistence proved to be their downfall when it was realised that they accumulated in biological systems.

The search for pesticides that were not so persistent but still effective was a challenge. Nevertheless, it was noted at the start of the millennium (Su and Scheffrahn 2000) that the presence of a single species of termite in the USA (*Coptotermes formosanus*) was sufficient to support a multimillion-dollar control industry with sales figures for liquid termiticide of $1.5 billion dollars annually. The industry is self-sustaining, because a termite that may be present in populations of several hundred thousand, foraging over a wide area, is going to be almost impossible to eradicate. Even if eradication seems to be successful, assurances cannot be given that the termites will not eventually return. There are plenty more where those came from.

In contrast, many other countries where termites invade buildings do not have a dedicated control industry, and repair, replacement or rebuilding is the accepted response to severe damage. Although this approach runs counter to the principles of heritage protection, resources may not be available for alternative technologies.

There is now an accelerating requirement for pesticides that have a minimal impact upon the environment. A 'green' treatment will be more suitable even though the consequences for the termites is less certain.

This has all generated a scientific literature that is too extensive to summarise. Therefore, this chapter attempts to draw together useful products and ideas that have emerged, but it cannot be comprehensive. Much has been developed for buildings with thin timber-framed walls, and it takes ingenuity to adapt these ideas to buildings that might have thick stone walls potentially containing timber, or the other variable constructions found in heritage buildings. The following sections discuss prevention and control strategies that have been developed, so that any that seem appropriate in a particular situation can be adopted or adapted.

7.1.1 Organochlorines and their successors

Wartime necessity promoted the development of synthetic insecticides during the 1940s. These organochlorine compounds included DDT, dieldrin and the termiticide soil poisons chlordane and heptachlor. They were inexpensive, had a broad spectrum of activity and seemed to be far less dangerous to use than some earlier choices such as arsenic. Unfortunately, their persistence in the environment, which made them excellent for uses such as timber treatment, also led to their residues accumulating in food chains. Then, in 1962, Rachel Carson's influential book, *Silent Spring*, alerted the world to the hazards of a careless reliance on pesticides.

This all prompted a search for more environmentally friendly products and in the 1970s and 1980s synthetic pyrethroids were produced. These were modelled on natural plant insecticides and were therefore deemed to be more suitable. Pyrethroids kill termites, but they also act as repellents, so the insects will avoid them and find an alternative route into a building. The toxicity of pyrethroids is also dose dependent, so if there is not enough to kill the termites they will probably recover. Moreover, pyrethroids are rapidly degraded by sunlight and are less persistent in soil than the organochlorines.

Organophosphates, based on phosphoric or phosphonic acid, such as chlorpyrifos, have also been developed and these make effective termite nerve poisons. But they have a short soil longevity and also have an adverse environmental impact.

A termiticide with a rapid 'knock-down' effect will kill the insects that it makes contact with, but might only have a limited effect upon a large colony. A termiticide that is slow acting and carried around the colony is ultimately likely to be more effective. A new range of termiticides became available during the 1980s and 1990s which were non-repellent and had a delayed action so that the poison could be passed from termite to termite. These pesticides have been formulated as liquids, foams, granules and dusts.

The following active ingredients are available (properties from the University of Hertfordshire PPDB: Pesticide Properties Database):

Imidacloprid: a neonicotinoid (synthetic nicotine) nerve poison that is effective even in small sublethal doses. It makes the termites inactive and unable to support the colony which in consequence becomes non-functional and perhaps destroyed by pathogens. Its half-life (50% remaining) in aerobic soil is about 191 days. It is highly water soluble (610mg/l @ 20°C) and is easily leached out by moisture. The ingredient is available in a foam formulation, which might be useful where there are cavities. Unfortunately, imidacloprid is highly toxic to non-targeted insects such as honeybees, as well as to birds and aquatic life.

Fipronil: a phenylpyrazole nerve poison formulated in a range of forms for agricultural and domestic insect pests, although toxic residues and toxicity to non-targeted insects such as honeybees have restricted its use in several countries. It is broken down by sunlight and has a half-life in aerobic soils of 142 days. Solubility in water is low (3.78mg/l @ 20°C). Its advantage for termite control is that it is passed from termite to termite by contact and trophallaxis

Chlorfenapyr: a halogenated pyrrole classed as a 'pro-insecticide' because it remains inactive until it enters the insect and is metabolised into an insecticide. It halts the production of energy, leading to cell dysfunction and death. Chlorfenapyr breaks down rapidly in aerobic soil and has a half-life of 1.4 days. It has low water solubility (3.78mg/l

@ 20°C) and low leachability. It is passed from termite to termite by contact and trophallaxis. Chlorfenapyr has low toxicity to mammals and is classified as a slightly hazardous insecticide by WHO criteria.

7.1.2 Plant extract termiticides

Confidence in available liquid termiticides has diminished. This is not because they are ineffective, but because of environmental concerns, including health hazards, their effects on non-targeted organisms and toxic residues. There is also the problem of increasing cost. Research on natural substances that act as termiticides has been progressing since at least the 1930s (Trikojus 1935). Extracts from plants and wood known to deter termites may be used as antifeedants (natural substances that will harm any creatures that eat them), deterrents or poisons. There are thought to be more than 2,000 plants belonging to 60 families that exhibit insecticidal properties (Dev and Koul 1997).

Table 7.1 lists some plant extracts that have been found to have an effect upon termites, mostly in laboratory assays. This list is not intended to be comprehensive. It is offered to demonstrate possibilities for termite control (*see* sections 7.2.2 and 7.2.4). If a plant or wood is known to be avoided by termites, then it is possible that an extract may prove useful. However, isolating and using the major active ingredient in these extracts may produce disappointing results because several ingredients may have a synergistic effect. Their extractive content and resultant effects on termites may also vary according to where and how the plant was grown. There is considerable scope for experimentation.

7.1.3 Boron compounds

Boron is a common element that in trace quantities is essential for metabolism in plants and animals (Gentz and Grace 2006). However, concentrations above those usually found in nature make boron compounds toxic to insects and fungi. However, excess boron in vertebrates is rapidly excreted by the kidneys, thus potentially making these compounds environmentally acceptable as pesticides.

Boron, particularly in the form of solutions of disodium octaborate tetrahydrate (DOT), has been used for many years as a preservative treatment for wood. Penetration for superficial application has been improved by using ethylene glycol as a solvent. The main problem has been found to be the slow leaching of the compound in some formulations.

Termites are not repelled by borates, but there does seem to be delayed avoidance (Grace and Campora 2005). These authors found that *Coptotermes formosanus* workers gradually ceased to forage near boron-treated wood over a few days, even though there were no dead termites present.

Borates are also effective when applied as dusts. One laboratory investigation (Green *et al* 2008) found that borax (sodium tetraborate) and zinc borate produced 100% mortality in seven and ten days respectively when dusted on to *Reticulitermes flavus*. The chemicals were also distributed around the colony by grooming, trophallaxis and cannibalism. Ingestion of the dusts killed the hind gut protists.

Table 7.1 Some plant extracts that have termicidal properties

English Name	Scientific name	Termite	Extract	Effect	Author
Tobacco	*Nicotiana tobacum*	*Microtermes* spp.	Aqueous of leaves (25gm powder in 100ml H_2O)	100% mortality in 24 hours	Shiberu *et al* 2013
Birbira	*Militia feruginea*	*Microtermes* spp.	Aqueous of seeds (25gm powder in 100ml H_2O)	100% mortality in 24 hours	Shiberu *et al* 2013
Endod	*Phylolacca dodecandra*	*Microtermes* spp.	Aqueous of seeds (25gm powder in 100ml H_2O)	100% mortality in 24 hours	Shiberu *et al* 2013
Caraway	*Carum carvi*	*Psammotermes hybostoma*	Oil (2ml on damp card)	Repellent	Aly *et al* 2012
Basil	*Ocimum basilicum*	*P. hybostoma*	Oil (2ml on damp card)	Repellent	Aly *et al* 2012
Camphor	*Eucalyptus globules*	*P. hybostoma*	Oil (2ml on damp card)	Repellent	Aly *et al* 2012
Clove	*Syzygium aromaticum*	*Coptotermes formosanus*	Bud oil (50µm/cm²)	100% mortality in 2 days	Zhu *et al* 2001
Vetiver	*Chrysopogon zizanioides*	*C. formosanus*	Root oil (25µm/g sand)	Repellent, halted tunnelling	Zhu *et al* 2001
Curry leaf plant	*Murraya koenigii*	*C. gestroi*	Crude leaf (10mg/ml)	Increased mortality with continuous exposure	Muda *et al* 2018
Green Chireta	*Andrographis paniculata*	*C. gestroi* and *Globitermes sulphureus*	Methanol extract of leaf (10,000ppm)	High mortality and highly repellent	Bakaruddin *et al* 2018
Neem	*Azadirachta indica*	*C. gestroi* and *Globitermes sulphureus*	Methanol extract of leaf (10,000ppm)	High mortality and highly repellent	Bakaruddin *et al* 2018
Tamarind	*Leucaena leucocephala*	*C. gestroi* and *Globitermes sulphureus*	Methanol extract of leaf (10,000ppm)	High mortality and highly repellent	Bakaruddin *et al* 2018
Gale of the wind	*Phyllanthus niruri*	*C. gestroi* and *Globitermes sulphureus*	Methanol extract of leaf (10,000ppm)	High mortality and highly repellent	Bakaruddin *et al* 2018
Mexican marigold	*Tagetes erecta*	*Odontotermes obesus*	Oil (6.3µl/cm²)	Good repellent	Verma *et al* 2016
Orange	*Citrus sinensis*	*O. obesus*	Oil (6.3µl/cm²)	Good repellent	Verma *et al* 2016
Orange	*Citrus* spp.	*C. formosanus*	Citrus peel oil	Fumigant for small enclosed spaces	Raina *et al* 2007
Japanese* cedar	*Cryptomeria japonica*	*Reticulitermcs speratus*	Wood vinegar (0.1ml on filter paper)	100% mortality in 1–3 days	Yatagai *et al* 2002

*Wood vinegar prepared with mixed chips also containing *Pseudotsuga menziesii* or *Quercus serrata* or *Pinus densiflora*.

7.1.4 Spinosad

This is an insect-control product produced by Dow AgroSciences. It is based on active ingredients produced by the actinomycete fungus *Saccharopolyspora spinosa* under aerobic fermentation conditions. Spinosad is a slow-acting and non-repellent toxin that is spread around a termite colony as individuals forage through residues. It has been found to be effective against *Cryptotermes brevis* (Woodrow *et al* 2006) when injected into galleries, and against *Coptotermes formosanus* when used in sand (Bhatta *et al* 2016). The later investigation found that three commercially available products mixed with sand at doses of 25ppm and 50ppm caused >85% mortality in one day and 100% in seven days.

7.2 Treatment of drywood termites

Drywood termites enter a building as alates or are imported in timber items. *Cryptotermes* are particularly easy to import in furniture (Hickin 1971). Termite alates will fly in through open windows and doors or enter via open access points, such as roof eaves. They live in small colonies, but these may be numerous and many may be inaccessible. The following sections summarise the treatment options that are currently available.

7.2.1 Fumigation

Whole-building fumigation with toxic gases has been used to control extensive drywood termite infestation in the USA since the early decades of the 20th century. The building is shrouded in a tent of gas-proof sheets or tarpaulins (Fig 7.1) and junctions between sheets are rolled together and secured with strong steel clips. The fumigant is introduced and then left for 24 hours so that it penetrates the timber and kills all the termites.

Early fumigants were hydrogen cyanide (HCN) and acrylonitrile (C_3H_3N) mixed with chloroform and carbon tetrachloride. These were replaced by methyl bromide (CH_3Br), but concerns about atmospheric

Fig 7.1
Tenting prior to fumigation.
[Rentokil archive photograph
© Rentokil Initial plc]

ozone depletion led to the fumigant being changed to sulphuryl fluoride (SO_2F_2). There is a positive synergistic effect if carbon dioxide is added.

These fumigants are highly toxic, but no residual pesticides are left behind. In consequence, reinfestation can commence soon after treatment. Ebeling and Wagner (1964) found that 26–37% of structures fumigated with methyl bromide in Los Angeles, USA, showed evidence of active termite infestation within three to five years. However, drywood termite infestations take many years to build up, so vigilance and localised treatment may help to control potential long-term problems.

7.2.2 Liquid termiticide treatments

When small colonies are located, they can be treated with liquid termiticide. Surface application is unlikely to penetrate more than a few millimetres, and then only if impermeable surface coatings have been removed.

This problem may be overcome by drilling and injecting the termiticide into termite galleries. A non-repellent termiticide, in some form, is likely to be most effective.

Orange-peel oil containing a high percentage of d-limonene has become a popular treatment in the USA (Mashek and Quarles 2008). It is produced by cold pressing the peel or steam distillation. The oil has a low environmental impact and the residual smell of oranges is generally acceptable, although the oil itself is classified as an irritant. Its major disadvantage is that there is no residual effect so the treatment must be targeted accurately – if galleries containing termites are not injected then the termites will survive. Although there may be a small fumigant effect, it is likely to be localised. Therefore, this type of treatment relies on termite colonies being locatable and accessible.

7.2.3 Desiccant and abrasive dusts

Termites are small creatures with a mostly soft and unsclerotised cuticle. Water retention to avoid desiccation is vital for their survival. The outer layer of the insect's body (epicuticle) is coated in a thin layer of wax about $0.1\mu m$–$1.0\mu m$ thick, formed from a complex mixture of lipids. Differences in its composition are important as non-volatile communication signals between insects, but the coating also controls moisture loss. A study of water loss in four species of subterranean and drywood termites (Zukowski and Su 2020) concluded that the thickness and composition of the wax layer was probably a better predictor of desiccation resistance than cuticle thickness. If the wax layer can be disrupted, then moisture loss would be difficult to control and the insect would die.

Early studies demonstrated that chemically inert dusts could be harmful to insects by promoting desiccation. Beament (1945) found that these dusts did not produce moisture loss in dead or stationary insects. The conclusion was that desiccation would be caused by abrasion of the wax layer as the insects moved around. Subsequent studies by Ebeling and Wagner (1959) found that sorbtion of the wax layer by some dusts would be more important. These authors showed that if alate drywood termites crawled around on an appropriate sorbtive dust, then it removed the wax layer, causing the termites to lose about 30% of their water content in one to two hours and die. The most promising dusts were silica aerogels. The effect was enhanced by soluble fluorides that had been added during the manufacturing process. They found that the gel adsorbed the wax layer and the fluorides then acted as a contact poison. The strong positive electrostatic charge imparted to the dust by the fluorides enhanced its ability to stick to the insects and the dusted surface.

Studies of over 200 inert dusts against a range of arthropods (Tarshis 1967) confirmed that synthetic silica aerogel containing ammonium fluorosilicate was the most effective. This was marketed as SG67, a flatting agent used to reduce the gloss of paints and varnishes.

Silica aerogel was an effective sorbtive dust because it had a high specific surface area (total surface area per unit mass) and a pore structure large enough to admit the molecules of cuticular wax. These gels are amorphous (non-crystalline) which greatly reduces their hazard to people and their LD_{50} (the dose given at once that causes the death of 50% of the test animal) is about the same as table salt. A formulation was first marketed in the USA in 1956 under the brand name 'Dri-Die'.

Other dusts were tested, and one called 'diatomite' became popular. Diatomite is formed from diatoms – the fossilised silicious remains of single-celled aquatic algae. This dust is much harder than silica gel and its effect seems to be more abrasive than sorbtive (Ebeling 1971).

Laboratory and field tests against bed bugs (*Cimex*) were carried out by the University of Kentucky (Potter *et al* 2014) using commercially available dusts. Their tests demonstrated that silica gel was significantly more effective than diatomite. This conclusion should also be valid for termites, as bed bugs, like alate termites, are secretive crawling insects.

Dusting was customarily used as a whole-space precautionary treatment, particularly following fumigation (Ebeling 1971). However, both silica gel and diatomite dusts are irritants and represent a respiratory hazard unless used strictly in accordance with the pesticide manufacturers' instructions. It is now usual to apply them as a thin deposit in cavities and targeted locations where insects might gain access or hide. These dusts are non-volatile and should retain their efficacy until they are cleaned up or obscured by dust.

Attempts to use desiccant dusts against subterranean termites have proved disappointing (Grace and Yamamoto 1993). Desiccation requires a good vapour pressure gradient between the insect and the environment. Humid environments, as required by subterranean termites, are less conducive to moisture loss. And if desiccation is very slow, the insect may be able to repair the damage to its wax layer.

7.2.4 Heat and chill
The maximum temperatures that drywood termites can survive have been ascertained for several species. Forbes and Ebeling (1987) estimated that *Incisitermes minor* could tolerate 49°C for 33 minutes, but *Cryptotermes brevis* and *Incisitermes snyderi* could not tolerate 50°C for more than 15 minutes (Scheffrahn *et al* 1997) and *Incisitermes immigrans* were killed at 46°C after 20 minutes. This has suggested that heating the timber in all or part of a building would be an effective treatment.

A typical procedure would be to remove sensitive items from the building, seal doors and windows or create a tent with tarpaulins, and duct in hot air from blowers until surface temperature probes registered over 54°C. This temperature would be maintained for 35 minutes or more. Sometimes propane heaters and fans are used, but propane combustion produces water vapour, which can be a problem with delicate finishes. The entire process may take four or five hours

It is sometimes difficult to ensure that there are no cooler areas due to materials that act as a heat sink, such as concrete structural elements. These may allow localised termite survival. The problem might be overcome by increasing the heat, but this may exacerbate damage to sensitive materials. This problem has been addressed by injecting wintergreen essential oil (methyl salicylate) as a synergistic fumigant into termite galleries prior to heat treatment (Perry and Choe 2020). There does not seem to be any provision to control relative humidity, and thus material desiccation, as is usual in European wood-boring beetle treatments by heat.

Localised heat treatment of infestations with microwaves is also popular. Typically, a microwave-emitting device is pointed at the infested timber or wall for a few minutes. Microwaves produce extreme heat that denatures protein and disrupts membranes (Hall 1988; Locatelli and Traversa 1989).

Drywood termites can also be killed by cold. Small holes are drilled and liquid nitrogen is pumped into cavities so that the temperature drops to −30°C. Care must be taken in small confined spaces to ensure that the operative is not suffocated by oxygen displacement as the liquid evaporates, and fragile items, finishes or services such as water pipes may be frozen and damaged. Liquid nitrogen is probably best used to treat small areas (<1m²) and components that can be covered with an insulation blanket (Forbes and Ebeling 1986).

Devices have also been marketed that produce lethal doses of electricity.

All of these methods will probably be effective if the risks are carefully considered and they are targeted by understanding the distribution of an infestation.

7.3 Treatment of subterranean termites

In most buildings, infestation by subterranean termites is assumed to be caused by workers invading from the soil, but some buildings may also be vulnerable to invasion from the air by winged alates (*see* section 6.1.2).

7.3.1 Poison barriers
The predominance of subterranean invasion has meant that earlier remedial termite treatments were devised to poison the ground under and around the building. This used large volumes of chemicals, but organochlorine termiticides such as chlordane were inexpensive. They were also persistent – field tests showed that they remained effective for more than 40 years. Sales literature from a major remedial company in the 1970s illustrates the method:

> The soil around all walls has to be treated. Where soil is exposed along
> outside walls, it is a simple matter to dig a narrow trench and then flood

with a large volume of insecticidal fluid. The soil returned to the trench is also sprayed so that a complete barrier is produced.

Termite treatments have to be adapted according to the individual situation, but commonly the trench would be about 150mm wide and dug to about 50mm below the top of the footings (Fig 7.2). This would probably equate to a depth of around 250mm–350mm. Termiticide is generally applied at a rate of about $100l/m^3$ of soil removed.

Fig 7.2
Trenching and flooding to form a toxic barrier around the walls. A large volume of fluid is needed.
[Rentokil archive photograph © Rentokil Initial plc]

Fig 7.3
Drilling (a) and rodding (b) to treat the ground beneath.
[Rentokil archive photograph © Rentokil Initial plc]

Where there is concrete, either an apron around the walls or a floor slab, we have to make holes with a rotary hammer drill at intervals and inject fluid.

Holes would generally be drilled 75mm–100mm away from the wall at 250mm–300mm intervals with a 300mm × 12.7mm-diameter bit.

Injection into the soil below would generally be with a tube (Fig 7.3).

Walls are drilled and injected at regular intervals depending on the type of block in use. Termites often use hollow blocks to move from one part of the house to another.

Wood block flooring is drilled and injected, then finished by inserting dowel rods, which are sanded, stained and polished to match the rest of the floor.

The method uses an enormous volume of pesticide. It is still sometimes used with the repellent pyrethroid termiticides. A repellent termiticide discourages the termites, but they may find a different route into the building because it is almost impossible to produce an unbroken barrier by drilling and injecting. This is particularly true where cost constraints or environmental concerns limit the volume of fluid used. Theoretically, any foraging termites already above ground should desiccate because they are denied access to the soil, but in a complex building they may come from ancillary nests around roof faults or plumbing.

This type of treatment can be particularly challenging in heritage buildings. Methods of construction can vary from one part of the building to another, and building fabric may have been altered by past interventions or deterioration. The internal construction of a wall may be haphazard, and earth or mortar may have eroded away leaving voids. Investigations into moisture movement through the walls, floors and timbers of old buildings by Historic England have always demonstrated that these are seldom, if ever, homogeneous structures. Liquids will always find and follow the easiest pathways, particularly when injected under pressure.

Non-repellent termiticides (see section 7.1.1) would be more effective in heritage structures. However, they are more expensive and there are potential environmental hazards. For example, misapplication can result in the contamination of soil or water. This might suggest that their use should be limited to targeted treatments. However, they are extensively used by pest-control professionals in countries where there is an established and regulated remedial industry. If used correctly, there can be a substantial reduction in the amount of fluid required, compared to repellent formulations. This is because the termites are not forced to find alternative pathways, and carry the slow-acting poison around the colony.

Barrier treatments are generally considered to need repeating every five to eight years.

7.3.2 Physical barriers

Ebeling and Pence (1957) showed that sand particles which were too compact and small for the termites to push through, yet too large for them to move, made a useful barrier against the western subterranean termite (*Reticulitermes hesperus*). These particles were 0.85mm–2.36 mm in diameter.

Further tests by Su and Scheffrahn (1992) found that particles sieved to the ranges 2.00mm–2.36mm and 2.36mm–2.80mm reduced the burrowing abilities of both *Coptotermes formosanus* and *Reticulitermes flavipes* through a 5cm column of the sand.

Acda (2013) constructed an experimental house on the campus of the University of the Philippines where there were populations of *Coptotermes*, *Macrotermes*, *Nasutitermes* and *Microcerotermes*. The house was built on a 5cm-thick layer of sandy volcanic debris (lahar). This ash had been screened to 1.18mm–2.14mm. The layer was covered with concrete, but pipe holes filled with ash were left to simulate cracks and the floor was scattered with logs of a non-durable timber to act as a bait.

The set-up was monitored over a five-year period during which time there were no signs of termite activity in the house, even though surrounding fallen branches were attacked. One termite species, *Microcerotermes losbanosensis*, started to form foraging tubes across the concrete and onto the wall but these were removed. The barrier forced the termites to make visible tubes that could be easily disrupted.

Other materials that have been used to form barriers include crushed rock, crushed glass and stainless-steel mesh.

7.3.3 Control with fungi, nematodes and bacteria

The use of selected fungi, nematodes or bacteria is well established for controlling a wide variety of insect pests, but trials with subterranean termites have produced mixed results. Spores of pathogenic fungi including *Metarhizium anisopliae*, a frequently exploited species, have to germinate on the exterior of the insects. Unfortunately, these spores are usually removed by the colonies' defensive mutual grooming (Rosengaus and Traniello 2001).

Experiments with baits containing the phase before sporing (pre-sporulitic) of the fungus showed that they might be ingested and passed through the gut of the termite, but germination was supressed by the antifungal properties of the faeces (Coghlan 2004).

Pathogenic nematodes (round worms) are theoretically ideal for the biocontrol of termites, but again results from field trials have been mixed. There are numerous factors associated with the chemical and physical properties of the soil and other factors, which may affect the outcome (Gaugler 1988).

Termite nest material and gallery linings contain Actinobacteria, which provide a natural defence against other microbes. The use of pathogenic bacteria does not seem to have received much attention from researchers.

7.3.4 Baits

Liquid termiticides are used as a barrier to protect the building. Baits are used to manage or supress the termites in the soil that might otherwise forage into the building. This is a different approach that may be used in isolation or in conjunction with other methods. Several different types of bait are available, but they are all applied in a similar way.

A typical arrangement would consist of small, basket-like plastic bait stations embedded in the soil at about 3m intervals around a building, normally at a distance of 300mm–600mm from the wall. The stations (or 'bait monitors') are filled initially with untreated wood (Fig 7.4) and inspected at intervals of perhaps 1 to 3 months. If termites are found, a toxic bait is inserted. The termites feed on this and take it back to the colony where it is passed around, leading to colony decline and perhaps elimination. Bait stations may also be installed above ground in the building on infested timber or intercepting foraging tubes (Fig 7.5).

Baits are environmentally friendly because termites are targeted and the amount of chemical is very small. They do not contaminate water, and do avoid the need to drill an array of holes in the building. However, it is important to remember that subterranean bait stations do not attract termites and are only found if the foraging termites encounter them. Success is gauged by monitoring the suppression of termites in the bait stations. But this does not necessarily mean that the termite colony has

Fig 7.4

Perforated plastic bait station lined with blocks of wood and buried in the ground.

Fig 7.5

An above-ground bait box containing layers of cardboard. The author was informed that the active ingredient was sodium fluoride and that the boxes were found to be effective (Hôi An, Vietnam).

been eliminated – there may be other reasons (perhaps temperature or soil moisture content) why activity in the station has ceased. Therefore, seasonal environmental fluctuations must be considered when formulating the baiting programme. It is also possible that the termites found in the bait station are not the ones that are invading the building.

Baits seem to work well with the Rhinotermitidae, but have limited success against the higher termites. This is an important consideration in tropical countries where buildings may be infested with a mixture of several genera. For instance, baits may control the generally dominant *Coptotermes*, but this might allow other termites, such as *Globitermes*, that were previously of lesser importance, to increase in number or invade (Lee *et al* 2007).

Even within the Rhinotermitidae, there may be variability. Grace *et al* (2000) note that sulfluramid (see below) is ten times more toxic to *Coptotermes* than to *Reticulitermes*. This meant that it could be used at a lower concentration (100ppm), which reduced feeding avoidance (caused by sublethal doses), but increased the time taken to receive a lethal dose. These authors also noted that superficially similar cardboard bait matrices from different manufacturers might vary in their acceptability to the termites.

The active ingredients listed below are used in commercial baiting systems. These systems are an expensive option because of the cost of frequent monitoring by the remedial team.

Insect growth regulators (IGR)

Hexaflumuron, noviflumuron, lufenuron, triflumuron and diflubenzuron are benzoylurea chitin synthesis inhibitors (Vahabzadeh *et al* 2007). These cause the termites to moult early without a properly formed exoskeleton (Acda 2013) so that they cannot develop and support the colony. The chemical is passed around the colony by trophallaxis, causing it to become non-functional and eventually extinct. Cabrera and Thoms

(2006) mention a trial around the USA, from 1998 to 2000, in which 53 colonies of *Reticulitermes* spp. were eliminated with hexaflumuron in a mean of 205 days. Another 74 colonies were eliminated with noviflumuron in 107 days.

Energy production inhibitors

Sulfluramid is a synthetic sulfonomide that breaks down into perfluorooctane sulfonate (PFOS), a toxic, highly persistent and bio-accumulative pollutant. However, the concentration used in termite baits is extremely low. Termites stop feeding, become lethargic and eventually die.

Hydramethylnon is an amidinohydrazone that has the same effect on the termites as sulfluramid.

Energy production inhibitors are usually used in conjunction with other forms of treatment.

Case Study 6: Saunton, Devon, England

As mentioned in Chapter 1, a colony of the subterranean termite *Reticulitermes grassei* was found in Devon in 1994. The colony seemed to have been progressing for several years, and had extensively infested two timber-framed bungalows, outbuildings and soil within a site area of at least 75m × 35m (Fig 7.6). This was sufficiently worrying for the government to initiate a 12-year termite eradication programme, coordinated by the Building Research Establishment (BRE). An initial spray treatment using a permethrin-based insecticide had failed and an intensive baiting exercise was planned (Verkerk and Bravery 2004).

Fig 7.6
Termites in Devon – the ongoing monitoring zone included 29 properties.

This was seen as a good test of baiting efficacy because there were no other termite colonies in the vicinity to potentially confuse the results by invading the depleted territory.

An 'eradication zone' for ongoing monitoring and treatments was delineated. This included 29 properties within an area of 500m radius. An inner circle of 75m radius was the treatment zone containing monitoring devices and baits. From 75m–200m there was an intensive monitoring zone containing 1,000 pine (*Pinus sylvestris*) sapwood stakes at 3m intervals in lines 10m apart (drilled through concrete where necessary). The area from 200m–500m was a buffer zone containing a smaller number of wooden stakes than the monitoring zone.

Receptacles containing rolled corrugated cardboard were buried in the soil as a monitoring system together with 1m lengths of perforated PVC conduit containing timber slats.

Two commercially available baiting systems (Sentri Tech® and Sentricon® supplied by Dow Agrosciences), comprising 152 bait stations, were installed in July 1998. The bait used was a hexaflumuron-impregnated cellulose matrix. One month after installation 17% of the devices close to the buildings were found to contain termites, but when baits were added the termites were repelled. This problem was overcome by using a novel bait consisting of thin pine sapwood wafers (1mm–2mm thickness) separated by spacers and in various configurations. These were impregnated with 0.5% w/v hexaflumuron in an aqueous extract of the wood decay fungus *Gleophyllum trabeum*. (The attraction of termites to decay fungi is discussed in Chapter 4.)

The new baits caused the termite population to collapse in 1999 and the last signs of activity were found in August 2000. It was decided, however, that confidence in the elimination of the colony would not be possible until there had been no activity for ten years. Unfortunately, termites were found again in 2009 and 2010.

In 2014 one of the property owners applied for planning permission to demolish their bungalow and replace it with a two-storey building in termite-resistant materials (Government 2014). The application was contentious, and one of the perceived material planning issues was the risk of accidentally spreading the termites outside the monitored area. Eventually, a planning inspector was appointed by the Secretary of State for Communities and Local Government and a hearing was held in January 2014 attended by the various interested parties. The BRE reported that no termites had been found for three years, but reiterated that elimination of the colony could not be accepted until ten years had elapsed.

Both the BRE and the inspector believed that the risk of spread could be managed by having two members of the BRE on site during the demolition and construction works. Furthermore, this would enable them to see if there actually were any termites remaining. The risk from carrying termites in infested soil was minimal because there was already a Forestry Commission Restriction of Movement Order forbidding the removal of soil from the site. However, the Secretary of State took the view that the risk was unacceptable.

It was announced in 2021 that no further termite activity had been found during the twice-yearly monitoring inspections and the infestation was presumed extinct.

This case study demonstrates that termite baits can rapidly control a colony, but that extermination may be a protracted task. Absence of termites in the monitoring stations does not necessarily mean absence of termites in the monitored site. However, in many situations the rapid depletion of a population followed by monitoring and vigilance would be a useful outcome. It is also helpful to note the following, especially if resources are limited:

- The Building Research Establishment staff who monitored the situation were not professional termite exterminators. Members of a conservation project team, with some background knowledge of termites, could undertake experiments and monitoring themselves.
- Many of the monitoring stations were made from readily available materials, such as perforated PVC pipe.
- The attractant in the bait was a simple extract of a decay fungus. Becker (1969) discussed laboratory rearing methods for termites and noted that 'brown rot' decay fungi were particularly attractive to termites, but that some 'white rot' species were toxic. This probably needs site experimentation because he found that the type of fungus varied the response. A similar investigation, carried out by English Heritage in the 1990s, found that ether extracts from the oak rot fungus *Donkioporia expansa* attracted deathwatch beetle (*Xestobium rufovillosum*) when impregnated into plaster pellets (unpublished). Decay in timber caused by *D. expansa* is commonly associated with infestation by these beetles, and the closely related 'white rot' fungus genus *Phelinus* has been associated with termite damage.
- The amount of termiticide used in bait stations is very small compared to barrier treatments, thus minimising environmental impacts.

7.4 Prevention

Depriving pest organisms of the resources they need to encourage and sustain them is a cornerstone of an integrated pest management (IPM) approach. The need to remove sources of water and repair buildings with inherently durable or treated timber is commonly mentioned – but it is rarely emphasised. Integrated pest management has remained the province of the pest control and remedial industry, but there is not much profit in checking people's drains. This book has attempted to focus on preventative solutions to termite infestations, particularly as they relate to heritage buildings. Undoubtedly, conservation project teams around the world will have been diligent in protecting the buildings in their care, but the preventative approach is not commonplace. Therefore, it is hoped that those diligent conservators will find something of interest here.

It is deceptively easy to provide a list of suggested remedial measures that should limit the risk of termite invasion. Unfortunately, historic buildings – particularly large and complex ones – can present with a wide range of problems, some of which might be insoluble. The best advice may be to return the fabric and drainage arrangements to the architect or builder's original intentions where there is evidence that these performed well. Every building is different, so every solution will be site specific. That said, some general recommendations apply across the board:

- Undertake regular, thorough and systematic inspections for signs of termites.
- Check trees and fences that surround buildings for termite nests and activity. Ensure that there are no tree branches in contact with buildings (*see* Fig 7.7). Remove any branches that overhang buildings.
- Remove vegetation growing out of walls.
- Remove logs or tree stumps in the vicinity of the building (Fig 7.7).
- Remove debris, particularly stacked timber and firewood, from around walls (Fig 7.8 and Plate 20). If walls are free from vegetation and debris, there is a better chance of detecting evidence of foraging by termites.
- Reduce external ground levels where they have built up above the level of the internal floors. Ideally, the floor level should be at least 250mm above the ground level.
- Where practical, provide inspection hatches for floor and roof spaces and concealed voids. When floorboards are replaced, they should be fastened using screws, rather than nails, to allow for easy removal.
- Ensure that drainage arrangements are fully functional and that rainwater and condensate from air conditioning equipment is discharged safely, away from the building.
- Ensure that any faults that allow water into the building are repaired promptly, and that previous temporary repairs are made effective and permanent. Timber used in repairs should be durable or pretreated with preservative. Cut ends of pretreated timber should be brush treated with preservative before fabrication.
- Before planning treatment, ensure that there is a current active infestation – don't be misled by historical damage.
- Formulate treatment strategies based on each individual situation and the resources available.
- Keep a detailed record of all inspections and interventions.

The most important thing is to understand what the termites are actually doing – how they are foraging and what they are feeding on. This indicates the level of threat they pose to a building. There has been a tendency in the UK and elsewhere for 'standard' remedial treatments to be triggered by the identification (and sometimes misidentification) of a pest fungus or insect, particularly if a guarantee is required. However, standard treatments are invasive and destructive. While this approach may be suitable in some cases, if misapplied it can have devastating consequences for the building and the funds available to conserve it.

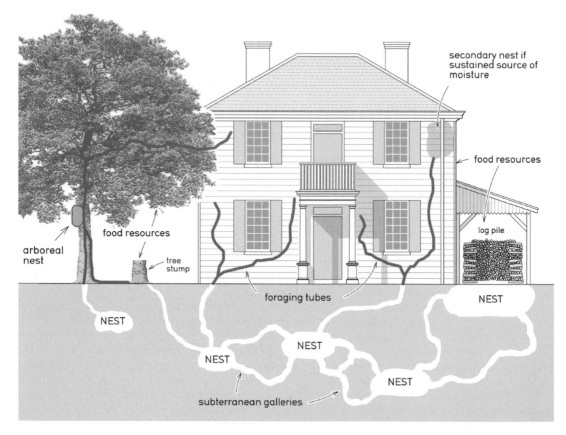

Fig 7.7
Subterranean Rhinotermitidae such as *Coptotermes* or *Reticulitermes* usually invade buildings from the ground, but may form satellite nests above ground if there is a persistent source of moisture. Arboreal termites such as *Nasutitermes* may invade perimeter building timbers via the ground, but might also use overhanging branches.
[© Iain McCaig]

Fig 7.8
Any timber debris that is removed from the building, even if intended to provide replacement components, should be stacked away from the walls. This stack was infested with *Coptotermes*.

Case Study 7: Timber-framed building, Turks and Caicos Islands

History

This building was constructed in 1815 using southern yellow pine (*Pinus* spp.) imported from the southern states of the USA (Fig 7.9). The timber, which was felled from natural forests, would have a good resistance to termites. The building had survived for 200 years.

Fig 7.9
A timber-framed building constructed in 1815 from yellow pine.

Fig 7.10
The termites had mostly attacked the pine at the interface between the boards and the frame, but unfortunately damage in some areas was more severe than hoped.

Fig 7.11
There was some termite damage in the roof timbers, but little termite activity was observed, suggesting that there had been a history of infestation at different periods in the building's history.

Termites

An infestation by the subterranean termite *Heterotermes tenuis* (Fig 7.10) was identified in 2015. *Heterotermes* termites are generally able to tolerate quite dry conditions and forage some distance away from their moisture resource. In this case they had formed foraging tubes in the central roof space and attacked roof timbers (Fig 7.11).

Sources of moisture

Some of the moisture required by these termites came from flashing and drainage faults at high level. Most of the moisture was beneath the ground floor where the soil was wet (Fig 7.12) in several areas. This was because of water penetration through open joints and cracks in the solid concrete veranda floor (Fig 7.13), together with a leaking rainwater disposal and collection system where water was diverted into a tank.

Fig 7.12
There was some termite activity under the ground floor, and damp soil indicated a current moisture problem.

Fig 7.13
The external balcony floor was cracked, with open joints that allowed water under the building.

History of damage

When the timber frame was exposed, the damage proved to be more extensive than expected. The termites responsible had been identified from photographs sent to the author in 2015 and so there was some level of ongoing infestation that needed to be controlled by removing the sources of moisture. However, this seemed to be localised, because no evidence of an active infestation was found when areas of boarding were removed during a site inspection that year. The point to be made here is that a termite infestation may flourish for a while when maintenance is poor and then die back or die out when it is improved. A large heritage building will have had a varied and unknown maintenance history. The extent of damage that is exposed during restoration works does not necessarily reflect the extent and severity of current infestation.

References

Acda, M N 2013 'Evaluation of Lahar barrier to protect wood structures from Philippine subterranean termites'. *Philipp J Sci* **142** (1), 21–5

Aly, M Z, Osman, K S, Mohanny, K M and Abd Elatti, Z A 2012 'Indoor and outdoor controlling evaluation on the subterranean termite, *Psammotermes hybostoma* (Isoptera: Rhinotermitidae) using some unordinary natural oils and others'. *Egypt Acad J Biol Sci. A, Entomol* **5** (2), 175–89

Bakaruddin, N H, Dieng, H, Sulaiman, S F and Ab Majid, A H 2018 'Evaluation of the toxicity and repellency of tropical plant extract against subterranean termites, *Globitermes sulphureus* and *Coptotermes gestroi*'. *Inf Process Agric* **5** (3), 298–307

Beament, J W L 1945 'The cuticular lipoids of insects' *J Exp Biol* **21**, 115–31

Becker, G 1969 'Rearing of termites and testing methods used in the laboratory' *in* Krishna, K and Weesner, F M (eds) *Biology of Termites*. New York and London: Academic Press, vol 1, 351–79

Bhatta, D, Henderson, G and Gautam, B K 2016 'Toxicity and non repellency of Spinosad and Spinetoram on Formosan subterranean termites (Isoptera: Rhinotermitidae)'. *J Econ Entomol* **109** (3), 1341–9

Cabrera, B J and Thoms, E M 2006 'Versatility of baits containing noviflumuron for control of structural infestations of Formosan subterranean termites (Isoptera: Rhinotermitidae)'. *Fla Entomol* **89** (1), 20–31

Carson, R 1962 *Silent Spring*. Boston: Houghton Mifflin

Coghlan, A 2004 'Green pesticide is irresistible to ants'. *New Sci* **184** (2476), 26

Dev, S and Koul, O 1997 *Insecticides of Natural Origin*. Amsterdam: Academic

Ebeling, W 1971 'Sorbtive dusts for pest control'. *Ann Rev Entomol* **16**, 123–48

Ebeling, W and Pence, R J 1957 'Relation of particle size to penetration of subterranean termites through barriers of sand or cinders'. *J Econ Entomol* **50**, 690–2

Ebeling, W and Wagner, R E 1959 'Rapid desiccation of drywood termites with inert sorptive dusts and other substances'. *J Econ Entomol* **52** (2), 190–207

Ebeling, W and Wagner, R E 1964 'Built in pest control'. *Pest Control* **32** (2), 53–62

Forbes, C A and Ebeling, W 1986 'Update: Liquid nitrogen controls drywood termites'. *IPM Pract* **8** (8), 1–4

Forbes, C A and Ebeling, W 1987 'Update: Use of heat for elimination of structural pests'. *IPM Pract* **9** (8), 1–5

Gaugler, R 1988 'Ecological considerations in the biological control of soil inhabiting insects with entomopathogenic nematodes'. *Agric Ecosyst Environ* **24** (1–3), 351–60

Gentz, M C and Grace, J K 2006 'A review of boron toxicity in insects with an emphasis on termites'. *J Agric Urban Entomol* **23** (4), 201–7

Government 2014 https://assets.publishing.service.gov.uk/government/uploads/system/uploads/attachment_data/file/319961/Called-in_decision_-_Brackens__Saunton__Braunton__ref_2201290__12_June_2014_.pdf

Grace, J K and Campora, C E 2005 'Food location and discrimination by subterranean termites (Isoptera: Rinotermitidae)' in Lee, C-Y and Robinson, W H (eds) *Proceedings of the 5th International Conference on Urban Pests*. Singapore: Executive Committee of the International Conference on Urban Pests, 437–41

Grace, J K and Yamamoto, R T 1993 'Diatomaceous earth is not a barrier to Formosan subterranean termites (Isoptera: Rhinotermitidae)'. *Sociobiol* **23** (1), 25–30

Grace, J K, Yamamoto, R T and Tome, C H 2000 'Toxicity of sulfluramid to *Coptotermes formosanus* (Isoptera: Rhinotermitidae)'. *Sociobiol* **35** (3), 457–66

Green, F, Arango, R A and Esenther, G 2008 'Transfer of termicidal dust compounds and their effects on symbiotic protozoa of *Reticulitermes flavipes* (Kollar)'. IRG/WP 08-10661, 1–9

Hall, D W 1988 'Pest control in herbaria'. *Taxon* **37** (4), 885–907

Hickin, N E 1971 *Termites: A World Problem*. London: Hutchinson

Lee, C-Y, Vongkaluang, C and Lenz, M 2007 'Challenges to subterranean termite management of multi-genera faunas in Southeast Asia and Australia'. *Sociobiol* **50** (1), 213–21

Locatelli, D P and Traversa, S 1989 'Microwaves in the control of rice infestations'. *Ital J Food Sci* **2**, 53–62

Mashek, B and Quarles, W 2008 'Orange oil for drywood termites: Magic or marketing madness'. *IPM Pract* **30** (1/2), 1–9

Muda, S M, Kamarozaman, A S, Mohamad, A, Ibrahim, M A N, Zani, A M and Muhammud, A 2018 'The bio-efficacy of crude leaf extract (Murraya koenigii) as botanical termiticides against subterranean termite *Coptotermes gestroi*'. *Int J Eng Technol* **7** (4.42), 78–80

Perry, D and Choe, D-H 2020 'Volatile essential oils can be used to improve the efficacy of heat treatments targeting the western drywood termite: Evidence from simulated whole house heat treatment trials'. *J Econ Entomol* **113** (5), 2448–57

Potter, M F, Haynes, K F, Gordon, J R, Washburn, L, Washburn, M and Hardin, T 2014 'Silica gel a better bed bug desiccant', www.pctonline.com, August, 77–84

Raina, A, Bland, J, Doolittle, M, Lax, A, Boopathy, R and Folkins, M 2007 'Effect of orange oil extract on the Formosan subterranean termite (Isoptera: Rhinotermitidae)'. *J Econ Entomol* **100** (3), 880–5

Rosengaus, R B and Traniello, J F 2001 'Disease susceptibility and the adaptive nature of colony demography in the dampwood termite *Zootermopsis angusticollis*'. *Behav Ecol Sociobiol* **50** (6), 546–56

Scheffrahn, R H, Wheeler, G S and Su, N-Y 1997 'Heat tolerance of structure infesting drywood termites (Isoptera: Kalotermitidae) of Florida'. *Sociobiol* **29** (3), 237–45

Shiberu, T, Ashagre, H and Negeri, M 2013 'Laboratory evaluation of different botanicals for the control of termite, *Microtermes spp* (Isoptera: Termitidae)', https://www.omicsonline.org/scientific-reports/2157-7471-SR-696.pdf

Su, N-Y and Scheffrahn, R H 1992 'Penetration of sized-particle barriers by field populations of subterranean termites (Isoptera: Rhinotermitidae)'. *J Econ Entomol* **85** (6), 2276–8

Su, N-Y and Scheffrahn, R H 2000 'Termites as pests of buildings' *in* Abe, T, Bignell, D E and Higashi, M (eds) *Termites: Evolution, Sociability, Symbioses, Ecology*. Dordrecht: Kluwer Academic, 437–53

Tarshis, J B 1967 'Silica aerogel insecticides for the prevention and control of arthropods of medical and veterinary importance'. *Angew Parasitol*, **8** (4), 210–37

Trikojus, V M 1935 'Some synthetic and natural antitermitic substances'. *Aust Chem Inst J Proc* **2**, 171–6

Vahabzadeh, R D, Gold, R E and Austin, J W 2007 'Effects of four chitin synthesis inhibitors on feeding and mortality of the eastern subterranean termite, *Reticulitermes flavipes* Kollar (Isoptera: Rhinotermitidae)'. *Sociobiol* **50** (3), 833–59

Verkerk, R H J and Bravery, A F 2004 'A case study from the UK of possible successful eradication of *Reticulitermes grassei*'. Final Workshop COST Action E22 'Environmental Optimisation of Wood Protection', Lisbon, Portugal, 22–23 March 2004

Verma, S, Sharma, S and Malik, A 2016 'Termicidal and repellency efficacy of botanicals against *Odontotermes obesus*'. *Int J Res Biosci* **5** (2), 52–9

Woodrow, R J, Grace, J K and Oshiro, R J 2006 'Comparison of localized injections of spinosad and selected insecticides for the control of *Cryptotermes brevis* (Isoptera: Kalotermitidae) in naturally infested structural mesocosms'. *J Econ Entomol* **99** (4), 1354–62

Yatagai, M, Nishimoto, M, Hori, K, Ohira, T and Shibata, A 2002 'Termicidal activity of wood vinegar its components and their homologues'. *J Wood Sci* **48**, 338–42

Zhu, B C R, Henderson, G, Chen, F, Fei, H and Laine, R A 2001 'Evaluation of vetiver oil and seven insect-active essential oils against the Formosan subterranean termite'. *J Chem Ecolog* **27** (8), 1617–25

Zukowski, J and Su, N-Y 2020 'A comparison of morphology among four termite species with different moisture requirements'. *Insects* **11** (5), 262, https://doi.org/10.3390/insects11050262

Glossary

Adultoid reproductives Primary reproductives (see below) that are retained within the nest they developed in and become functional (Termitidae).

Alates (from the Latin for wing) Primary reproductives (king or queen termites) that develop from nymphs. They grow wings and leave the nest in a swarm to form other colonies.

Carton Nest material made from faeces, partly digested wood and soil.

Chitin The second most plentiful polysaccharide after cellulose. It is a similar linear polymer but with the addition of acetyl glucose amine. It is found in the exoskeletons of arthropods including insects, and also in fungi and bacteria.

Cuticle The exoskeleton or protective 'skin' or 'shell' covering insects. Some parts are flexible, while others are hardened plates for protection (see **Sclerotised**). There is a thin outer layer of wax for waterproofing.

Embolism The blocking of water conduction through vessels and tracheids by air bubbles.

Epigeal A nest constructed partially below and partially above ground.

Ergatoid reproductives Secondary reproductives that develop from workers.

Eusocial A level of sociality including cooperative brood care, overlapping generations and differentiation of roles into reproductive and non-reproductive groups.

Flagellate protozoa Single-celled organisms (protists) with one or more whip-like flagella that help with movement, attachment or feeding.

Fontanelle A pit between the antennae on the heads of some termites that is the opening for the frontal gland that produces defensive secretions. It is particularly prominent in the Rhinotermitidae.

Homeostasis The maintenance by self-regulation of conditions that are optimum for survival.

Humeral suture A line of weakness at the bases of the wings where they break off after flight.

Humification Gradations in the decomposition of organic materials in soils and peats.

Larvae Immature individuals without wing buds.

Mycelium An interwoven mass of fungal threads (hyphae).

Neotenic reproductives Secondary reproductives, which may be nymphoid or ergatoid.

Nymphoid reproductives Secondary reproductives that develop from nymphs.

Nymphs A development stage from larvae where wings are starting to form (wing buds).

Pheromones A chemical or mix of chemicals that is released to the exterior and causes a specific modification in behaviour in a member of the same species.

Phragmotic Phragmosis is the defence of an organism in its burrow by using its own body. Some *Cryptotermes* have phragmotic heads, which are plug-shaped to block galleries in the nest during attack by predators.

Physogastric Termite queens are physogastric. This means that their abdomen may swell with eggs to many times its original size.

Primary reproductives The king and queen that found a colony.

Propagule A part of a plant or fungus that can grow to produce a new individual.

Pseudergates False workers found in the Termopsidae and Kalotermitidae that retain the ability to turn into winged reproductives (alates), unlike true workers.

Sclerotised Hard, dark regions in the cuticle of insects that contain sclerotin, a material formed from cross-linked proteins by a process called sclerotisation – a form of tanning.

Secondary reproductives See **adultoid reproductives**, **ergatoid reproductives**, **neotenic reproductives** and **nymphoid reproductives**.

Soldier A caste with defensive adaptations such as enlarged and heavily sclerotised mandibles.

Supplementary reproductives Produced by the queen by parthenogenesis (absence of sexual reproduction) when more queens are required.

Symbiotic Close ecological relationship between two different organisms living together.

Synergy The interaction between two or more chemicals or organisms to produce an effect greater than if each were acting alone.

Tandem Alates lose their wings and run around in pairs looking for a new nest site. One, usually the male, hangs on to the abdomen of the other.

Tanned A tanning agent displaces water from between protein fibres and cements them together.

Tergite A dorsal thickened and hardened (sclerotised) plate on the abdomen of an insect.

Transpiration Water movement through a plant and its loss through evaporation through the aerial parts such as leaves. This produces mass flow of water and nutrients up the plant.

Trophallaxis The exchange of food and other fluids between social insects in a community.

Vascular pathogen A pathogen that blocks the functioning of the plant vascular system so that water transport is halted.

Venation Veins that strengthen the wings of insects. Their pattern and interconnections may be diagnostic for termite families.

Worker A sterile form without differentiation towards alates of soldiers that provides food and undertakes construction and repair.

Index

Plate 1
King and physogastric queen of *Reticulitermes hageni*. Increasing egg production stretches the soft connective tissue between the skeletal plates, eventually giving the queen a striped appearance. [© Lyle Buss, University of Florida]

Plate 2
Two preserved queens of *Coptotermes sp.* that have been dug from a nest. They are the same size as the two fingers visible on the left side of the bottle neck.

Plate 3
Eggs and larvae of *Reticulitermes flavipes* tended by a worker. The larvae are, in appearance, just smaller versions of the worker. [© Doug Wechsler]

Plate 4

Swarm of *Reticulitermes flavipes*. Some larvae will develop as nymphs with wing buds and eventually become alates. These are winged reproductives (future kings and queens) that will leave the nest in a swarm when population density and weather conditions are appropriate.
[© Lyle Buss, University of Florida]

Plate 5

A pile of shed wings, probably *Cryptotermes sp.*, that have been left when the paired kings and queens entered cracks to commence colonies in the basement of a monastery.

Plate 6

Workers of *Hypotermes xenotermitis* grazing on the softened surface of a teak beam. They are guarded by a fringe of soldiers.

Plate 7
A soldier of *Coptotermes formosanus*. Note the forward-pointing fontanelle on the top of the head. This is the aperture through which defensive secretions produced by the frontal gland are projected.
[© Lyle Buss, University of Florida]

Plate 8
A soldier of *Reticulitermes flavipes*. *Reticulitermes* is the termite genus that is successfully colonising southern Europe, and *R. flavipes* was first described from a palace hothouse near Vienna in 1837. It has proved to be impossible to eradicate in Hamburg where it infests the municipal heating system.
[© Lyle Buss, University of Florida]

Plate 9
Soldiers of *Kalotermes flavicollis*, the common drywood termite of the Mediterranean region. The soldiers have toothed mandibles, but these are difficult to see without a lens. Soldiers generally have a more cylindrical body shape than the *Reticulitermes* they might be confused with in Europe.

Plate 10
Drywood termites produce
seed-like faecal pellets that
are pushed out of the timber
and form distinctive mounds.

Plate 11
A *Kalotermes* colony within
a single piece of skirting
board. Part of the surface has
cracked into cubes because
of fungus damage (brown
rot) demonstrating that the
termite colony developed
where there was water
leaking into the building.

Plate 12
Subterranean termites colonise buildings via foraging tubes, which can provide both a localised damp environment and protection from predators.

Plate 13
The same material used to form tubes may also be used to line galleries in colonised timber.

Plate 14
Two sections from the bottom of base logs cut from 200-year-old Scots pine (*Pinus sylvestris*). The large section was grown in a park in the West Midlands, UK. The small section came from Kenozero in northern Russia. The short northern growing season restricts growth and explains the considerable difference in diameter.

Plate 15
A section of a 19th-century Scots pine joist, with about 130 annual growth rings, thus demonstrating that it was probably imported into the UK via the Baltic ports and originated in Russia or Poland. This is all heartwood, with a good resistance to wood-damaging fungi and insects, including termites. (The base fractured because of an internal growth defect, not decay.)

Plate 16

Two plantation-grown sections of Scots pine stained to show the development of heartwood. The white boxes represent timbers that could be cut with the same dimensions as the example in Plate 15. (A) is a section from the first commercial 'thinning' of the plantation at about 18 years. There is one small growth ring of heartwood developing at the centre, and timber cut from this trunk would be all perishable sapwood. (B) is taken from the second thinning at about 35 years. The heartwood is now noticeable, but the cut timbers would still be mostly sapwood. Original timbers should be conserved, not only because of their historic value, but because they often have a durability that cannot now be replicated except with additional treatments.

Plate 17

If a suitable food resource is located, then the tube material produced by some subterranean genera may be spread out as a sheet. In this photograph the termites under the sheet are grazing on the sapwood surface of mangrove ceiling poles.

Plate 18
The end of a mangrove ceiling pole. The unprotected sapwood has been removed, but the heartwood chemistry makes most of the timber resistant to termites.

Plate 19
The end of a nearby pole where a damp wall has allowed a fungus to colonise the bearing. The fungus has modified the chemistry of the heartwood and the modified section is now available for the termites.

Plate 20
If debris and plant growth is kept away from walls as far as is practical, then termite activity becomes easier to detect and a programme of inspection and repair makes established infestation less likely.